Epigenetics - Regulation and New Perspectives

Edited by Tao Huang

Published in London, United Kingdom

Epigenetics - Regulation and New Perspectives
http://dx.doi.org/10.5772/intechopen.104312
Edited by Tao Huang

Contributors
Shaohua Qi, Guofei Cui, Qing Dong, Kexin Gai, Denis Mustafov, Sara Seriah, Roozba Malik, Maria Braoudaki, Ewelina A. Klupczyńska, Sagrario Lobato, Andrea Pérez, Ivonne Ramírez, Alejandra Morales, Karla Rubio, Olalekan Bukunmi Ogunro, Oladimeji Taiwo Babatunde

First published in London, United Kingdom, 2023 by IntechOpen
IntechOpen is the global imprint of INTECHOPEN LIMITED, registered in England and Wales, registration number: 11086078, 5 Princes Gate Court, London, SW7 2QJ, United Kingdom

British Library Cataloguing-in-Publication Data
A catalogue record for this book is available from the British Library

Additional hard and PDF copies can be obtained from orders@intechopen.com

Epigenetics - Regulation and New Perspectives
Edited by Tao Huang
p. cm.
Print ISBN 978-1-83768-393-2
Online ISBN 978-1-83768-394-9
eBook (PDF) ISBN 978-1-83768-395-6

For EU product safety concerns:
IN TECH d.o.o., Prolaz Marije Krucifikse Kozulić 3, 51000 Rijeka, Croatia,
info@intechopen.com or visit our website at intechopen.com.

Meet the editor

Tao Huang is a professor at the Shanghai Institute of Nutrition and Health, Chinese Academy of Sciences. He completed his post-doctoral research at the Department of Genetics and Genomics Sciences, Icahn School of Medicine, Mount Sinai, New York City, USA. His research interests include bioinformatics, computational biology, systems genetics, and big data research. He has published more than 200 articles. He has been an editor or guest editor for more than 30 journals and a reviewer for 230 journals. He was recognized as a Highly Cited Chinese Researcher in 2020, 2021, and 2022 and listed among the World's Top 2% of Scientists in 2020, 2021, and 2022.

Contents

Preface

Epigenetics plays a crucial role in regulating gene expression, particularly in the field of transcriptomics. Dysfunctions in epigenetics have been identified as key factors in the development of cancer and numerous complex diseases. The ability to detect cancer at an early stage through the analysis of DNA methylation in blood samples has garnered significant interest in the field of liquid biopsy. This is because changes in DNA methylation patterns can be detected before changes in gene expression, making methylation analysis a more effective method for early cancer detection. By identifying signals of tumorigenesis, we can initiate treatment for cancer as early as possible, thereby greatly improving the prognosis for patients.

In addition to DNA methylation, there are numerous emerging areas of study in the field of epigenetics. These include research on the 3D genome, which is based on analyzing Hi-C data, as well as investigations into RNA modification through sequencing of N6-methyladenosine (m6A), N1-methyladenine (m1A), 5-methylcytidine (m5C), and more.

In this book, we present the most recent advancements in epigenetics omics technologies and their utilization in understanding the mechanisms of cancer and complex diseases. Our intention is to provide readers with a comprehensive understanding of these technologies, inspiring them to adopt a multi-omics data integration approach in order to unravel the intricate mechanisms of epigenetic regulation in various biomedical contexts. We hope that this book will serve as an enlightening resource, broadening the horizons of our readers and motivating them to explore the potential of integrating multi-omics data for investigating epigenetic regulations in different biomedical scenarios.

Tao Huang
Shanghai Institute of Nutrition and Health,
Chinese Academy of Sciences,
Shanghai, China

Chapter 1

Chromatin Dynamics: Chromatin Remodeler, Epigenetic Modification and Diseases

Guofei Cui, Qing Dong, Kexin Gai and Shaohua Qi

Abstract

The gene transcription patterns are regulated in response to extracellular stimuli and intracellular development programs. Recent studies have shown that chromatin dynamics which include nucleosome dynamics and histone modification play a crucial role in gene expression. Chromatin dynamic is regulated by chromatin modification enzymes including chromatin remodeling complex and histone posttranslational modifications. Multiple studies have shown that chromatin dynamics dysregulation and aberrant and histone modifications resulted in the occurrence of various diseases and cancers. Moreover, frequent mutations and chromosomal aberrations in the genes associated with subunits of the chromatin remodeling complexes have been detected in various cancer types. In this review, we highlight the current understanding of orchestration of nucleosome position, histone modification, and the importance of these properly regulated dynamics. We also discuss the consequences of aberrant chromatin dynamic which results in disease progression and provides insights for potential clinic applications.

Keywords: chromatin dynamic, chromatin remodeler, epigenetic modification, gene regulation, histone modifications

1. Introduction

In eukaryotic cells, chromatin is the genetic material carrier which packaged of DNA with histone and non-histone proteins. The simplest form of chromatin structure is the nucleosome core particle. Each nucleosome is composed of 147 bp of DNA wrapped around an octamer of histone proteins (two copies each of core histones H2A, H2B, H3, and H4) plus a linker histone (H1) involved in higher-order chromatin compaction [1]. In general, chromatin assembly limits the accessibility of genomic sequences, and thus it creates inherent barriers for nuclear events such as transcription, DNA replication, and DNA repair. Consequently, chromatin structure must be regulated dynamically, and its compaction and assembly are regulated by multiple mechanisms, including DNA methylation, histone post-translational modification, histone variant incorporation, chromatin remodeling, histone eviction, and non-coding RNA pathways [2–4]. During the DNA replication, DNA damage repair, and transcription process, the assembly and disassembly of chromatin structure stay on a dynamic and balanced status-first, the core histones in front of the replication fork or activated transcription region need to be released from the nucleosome to allow the DNA replication and RNA polymerase II machinery to passage and then reassemble again after these processes are completed, as the events of DNA replication and RNA transcription occurring, the chromatin assembly and reassembly also keep a dynamic and balanced process [5–8]. Meanwhile, the nucleosome position is not randomly arranged; instead, it is regulated by chromatin remodeling complex to either condense or loosen status at different loci of the nucleus. Chromatin is compacted into higher-order structures-named chromosomes, including loosely packaged euchromatin that is open and functionally active and more condense packaged status that is relatively repressive heterochromatin and maintains genomic integrity [9–11].

The chromatin remodeling complexes are the second major class of chromatin regulators. They are involved in different biological events such as DNA replication and transcription, through altering the components, positions, and numbers of nucleosomes around the gene. It also involves diverse modulators and protein domains for the various processes: nucleosome organization, disorganization, ejection, or changes in nucleosome composition [12, 13]. These chromatin remodeler complexes constitute a highly related family of multi-subunit complexes, and the core catalytic subunit is comprised of the ATPase domain that hydrolyzes ATP. Therefore, the chromatin remodeler complexes are ATP-dependent chromatin-remodeling enzymes that use the energy of ATP hydrolysis to remodel nucleosomes. Other substrate recognizing subunits direct the nucleosome sliding, facilitate the access of transcription factors to nucleosome DNA, change the DNA topology on specific nucleosomes targets, and generate distinct remodeling outcomes [14, 15].

Here, we focus on the studies on the related chromatin remodeling complexes and epigenetic modification and summarize recent advanced knowledge on the power of chromatin remodelers and its associated dynamically epigenetic regulation. Moreover, we explored the correlations between chromatin dynamic regulation and diseases progression, highlighted the importance of various chromatin modifiers targets for disease therapy. We also discuss emerging evidence of the new roles for chromatin regulators in developmental transitions in the future clinic application. Given that most knowledge about the chromatin remodeling complexes are described in yeast; therefore, the text below will be defined as yeast if there is no extra interpretation.

2. Chromatin remodeler

As we know, the position and status of chromatin structure are not permanently stable; conversely, it is dynamically regulated by chromatin remodeling complexes, by using the energy of ATP hydrolysis to create a force to promote the local repositioning of nucleosomes and alter the accessibility of DNA elements to transcription factors and (or) other proteins [16, 17]. The activity of the ATP-dependent chromatin remodelers includes the exchange of core histones/histone variants, the eviction of histones from nucleosomes, and the repositioning or sliding of nucleosomes along DNA [18–20].

All the ATP-dependent chromatin remodeling complexes are multi-subunit complexes containing an ATPase subunit of the Snf2 (sucrose non-fermenting 2)-type of helicase. Based on the structural characteristics of this catalytic ATPases subunit, the ATP-dependent chromatin-remodeling complexes can be classified into four subfamilies, including SWI/SNF (switch/sucrose non-fermentable), ISW1 (imitation switch), CHD (chromodomain-helicase DNA-binding protein), and INO80 (inositol requiring 80) (**Figure 1**) [21–27]. Currently, the most thoroughly studied remodelers are SWI/SNF subfamily [28, 29], which is defined by its N-terminal HSA (Helicase-SANT–associated) domain working for binding actin and other actin-related proteins [30]. The histone acetylated-lysine binding domain is located at C-terminal, named bromo domain [31]. This remodeler family is large, multi-subunit complex that contains more than eight proteins.

Here in **Table 1**, we summarized these four chromatin-remodeler families and their subunits, and other functional domains are also described.

The formation of nucleosome structure is the natural obstacle for the processes of DNA replication and transcription (*please keep in mind that nucleosome structure also protects the fragile DNA from insult). How does the cell work for transcription under the condition of chromatin? One of the mechanisms is histone exchange or removal from nucleosomes mediated by histone chaperon and/or chromatin remodeler. For example, when activated transcription occurs, chromatin remodelers such as the Remodels the Structure of Chromatin (RSC) complex maintain these nucleosome-depleted regions (NDRs) by sliding nucleosomes away from the promotor region, allowing the binding of RNA polymerase II to the promoter [32, 33]. In addition, chromatin remodeling complexes promote the binding of transcription activators on gene promoters or enhancers region, finally resulting in the gene activation. However, the formation of the NDR alone does not ensure Pol II-mediated transcription initiation on special gene locus. This indicates that other mechanisms may be involved in this process. For example, it is shown that histone variant H2A.Z is specially located on the two nucleosomes flanking the NDR (denoted as −1 and +1 with respect to the NDR) at certain genes, promoting or inhibiting gene expression. The incorporation and removal of H2A.Z into +1 and −1 nucleosomes are mediated by chromatin remodeling complex SWR and INO80 [34, 35].

Both the processes of DNA replication and RNA transcription involve nucleosome assembly and organization, the histone complexes (H3–H4 tetramers and H2A–H2B dimers) are delivered by histone chaperones for chromatin-remodeler (including ISWI and CHD subfamily). After DNA is exposed from histones, access subunit(s)

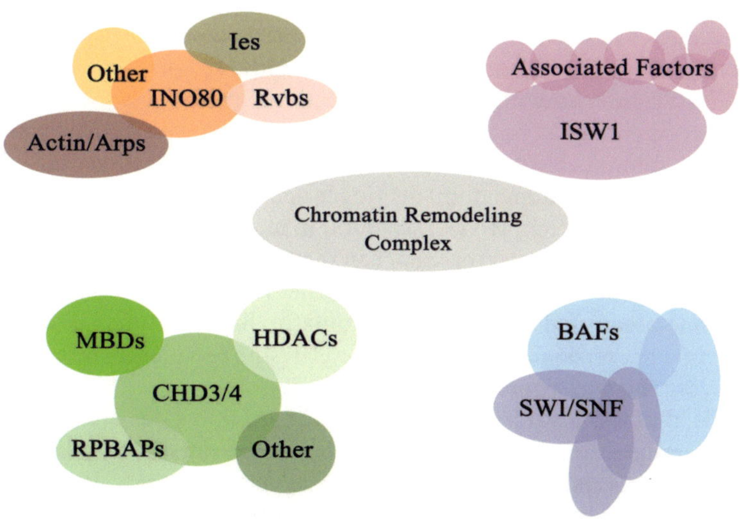

Figure 1.
Composition of the chromatin remodeler complexes.

Chromatin-remodeler families	Subunits	Domains	Functions
SWI/SNF	SWI1, SNF11, Swp82, SWI2/SNF2, Swp73, SWI3, Arp9, SNF5, Arp7, SNF6, Swp29	HSA	Actin binding
		DExx	ATPase
		HELICc	ATPase
		BROMO	Acetylated lysine binding
ISWI	ISW1, P74, p110, p105, ACF, RSF, CERF, CHRAC, NURF, NoRC, WICH, b-WICH	DExx	ATPase
		HELICc	ATPase
		SANT	Histone binding
		SLIDE	Histone binding
CHD	CHD1, CHD2, CHD3, CHD4, CHD9, NuRD	CHROMO	Histone binding
		DExx	ATPase
		HELICc	ATPase
INO80	Arp8, Arp4, Taf14, Ies4, Actin, Ino80, RvB1/2, Ies3, Ies1, Ies2, Arp5, Nhp10, Ies5, Ies6, SRCAP	HSA	Actin binding
		DExx	ATPase
		HELICc	ATPase

Table 1.
Four chromatin-remodeler families, subunits, and their respective functional domains.

of the remodelers will direct the binding of transcription activators or repressors on gene promoters or enhancers region. Meanwhile, remodelers also help to protect the "naked" DNA by recruiting other protein complexes, including the histone modification, transcription activator, or repressor complex, and during this process, the electrostatic environment and space surrounding the chromatin are also involved.

In addition, the nucleosome structure can be remodeled by the interaction between histones (variants) modifications and special recognizing subunits of the chromatin remodeler. The remodeler-specific domains recognizing and binding histone modifications, generally reference as reader domains. It is shown that the remodelers have a greater affinity for the nucleosome than naked DNA, which means the recognizing subunits have a priority to bind the modified histones [16]. For example, the PHD (plant homeodomain) finger domain, discovered over a decade ago in the Arabidopsis protein HAT3.1, is found in many chromatin-remodeling proteins. It functions as an "effector" that can recognize histone H3 tail peptides at lysine 4 (H3K4me2 and H3K4me3) [36–39], further recruiting transcription factors and nucleosome-associated complexes to chromatin. Bromo domain, another protein recognition module, recognizes and binds acetyl-lysine residues on histone tails protruding from the nucleosome [40, 41].

Meanwhile, the histone variants also influence the affinity of these remodeler subfamilies, and certain histone modification variants markers can recruit specific chromatin remodeling complexes and further reinforce the remodeling. During this process, the chromatin remodelers play an important role in the "position effect" of gene expression [42, 43]. There are also some other protein recognition modules that have been described in the past two decades; however, we will not describe them one by one here in this chapter.

3. Chromatin dynamics and histone modifications

Both histone tails and globular domains are subject to a vast array of different posttranslational modifications including acetylation, methylation, phosphorylation, deamination, β-N-acetylglucosamine, ADP ribosylation, (de)ubiquitylation, and SUMOylation [44]. Histone methylation frequently occurs at lysine (K) and arginine (R) residues, which is mediated by histone methyltransferase such as SET-1 (H3K4me), SET-2 (H3K36me), and PRMT5 (H3R8, H4R3); histone acetylation occurred at lysine residues is mediated by acetyltransferase, such as GCN5 (H3K9, K14, and K18) and HAT1 (H4K5 and K12). Phosphorylation occurred at serine and threonine residues is achieved by MSK1/2 (H3S28). Ubiquitination occurred at lysine residues is mediated by ubiquitinase such as RNF20/RNF40 (H2B120K). Histone modifications that are associated with active transcription are commonly referred to as euchromatin modifications, such as acetylation of histone 3 and histone 4 (H3 and H4) or di- or trimethylation of H3K4 [18, 44]. However, histone modifications that are occurred at inactive genes or regions are often termed heterochromatin modifications, such as H3K9me and H3K27me. There are two well characterized mechanisms for the function of histone modifications. The first is the disruption of contacts between nucleosomes and nucleosomes or DNA in order to "unravel" chromatin; the second is the recruitment of nonhistone proteins to bind to chromatin or to help remove histone/histone variant from chromatin [18]. Given the diversity of covalent modifications, it has been proposed that individual histone modifications or modification patterns might be read by other proteins that influence chromatin dynamics and function.

Many histone-modifying enzymes are components of chromatin remodeler complexes. For example, the Bdf1 subunit of chromatin remodeling complex SWR contains two bromodomains that bind to acetylated-lysine in H3 and H4 [45]. This indicates that the various kind of histone modifications cooperate with chromatin remodelers to modulate gene expression by altering the chromatin structure. From a genome-wide of view, histone H3K4me3 and H3ac are strongly correlated with active transcriptional start sites [46, 47]. Conversely, H3K9me3 are usually located on CpG island and mediates heterochromatin formation and gene silencing [48, 49].

In most cases, the chromatin-associated factors have been shown to specifically interact with modified histones rather than DNA, then chromatin remodelers are recruited to DNA locus independently. For the recruitment of chromatin remodeler

Chromatin remodeler readers/domains	Histone target(s)	Example
Bromodomain	Acetylation Lysine	H3K14ac [50]
Chromo domain	Methylation Lysine	H3K4me, H3K9me, H3K27me, H3K36me [51]
PWWP (Pro-Trp-Trp-Pro motif)	Methylation Lysine	H3K36me [52]
PHD finger	Methylation Lysine	H3K4me [53, 54]
MBT (malignant brain tumor) repeat	Methylation Lysine	H4K20me [55]
SANT domain	non-modified histone tails	[56, 57]

Table 2.
The major chromatin remodeler readers/domains and their binding histone targets.

complexes to chromatin, the transcription factors with distinct DNA binding domains work to direct the target selectivity and functional specificity.

Here, we summarize the chromatin remodeler readers and their histone targets with various modifications (**Table 2**).

In fact, all the epigenetic modifications cooperate with each other to guarantee an accurate regulation of gene expression. Certain histone modification markers can recruit specific chromatin remodeling complexes and further reinforce the remodeling of nucleosomes. During this process, the chromatin remodelers play an important role in the "position effect" of gene expression [16].

4. Crosstalk among chromatin dynamics, epigenetic modifications, and gene regulation

How to determine a gene's expression status? What extent should it be silenced or activated? The answer may depend on the chromatin position or the state of the target gene. Generally, each type of epigenetic modifications is dynamic and keeps on changing within the cell, and it is driven by cell signals induced by alterations in the cellular environment, including changes in nutrients, stress, hormone levels, and cell damage, etc. The cross-talk among chromatin dynamics, epigenetic modification, and gene expression regulation is an extra complex process via multiple possible mechanisms: (1) these events may be dependent on another; (2) they may work competitive; and (3) one factor disruption/mutation do not necessarily have

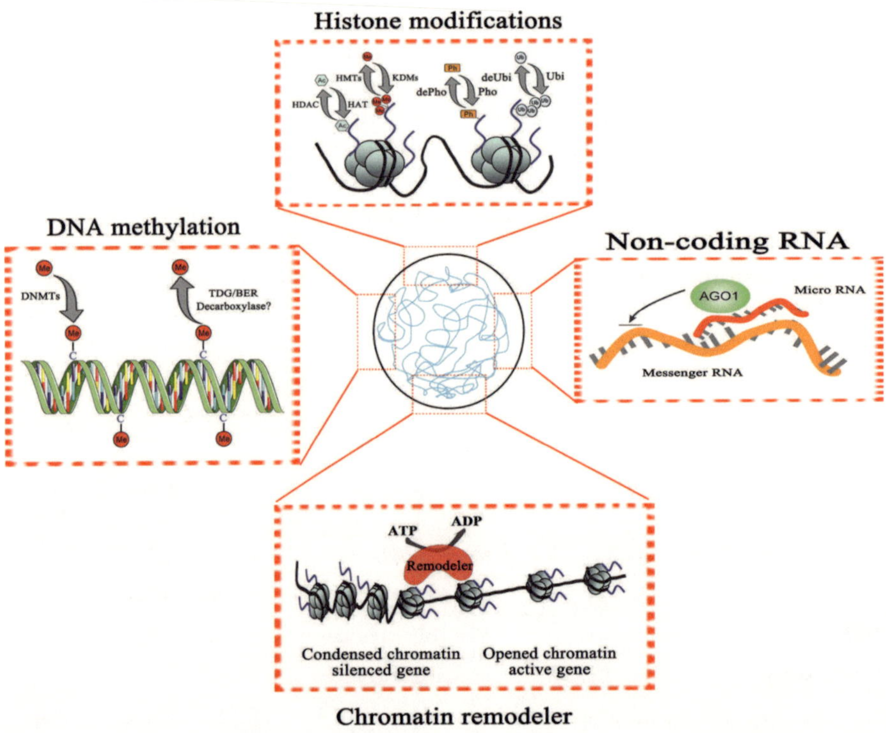

Figure 2.
Dynamic of epigenetic modifications.

effect on another directly. In addition, most of the genes, especially those involved in cell differentiation and proliferation, may have various layers of regulatory mechanisms [58]. For example, DNA methylation can reinforce histone modifications and strengthen chromatin structure, to ultimately activate or silence the expression of specific genes [59]. In some cases, they may function in competition to "check" each other to maintain homeostasis [60, 61]. A promotive histone modification marker and a repressive chromatin structure can co-exist to either keep gene expression at a moderate level or switch it "on and off" at different time points, quickly and efficiently [62, 63] (**Figure 2**). The histone modifications can direct the gene expression regulation by the chromatin remodeler. It was reported that H3K56Ac alters the substrate specificity of SWR-C, resulting the random switch of either H2A.Z/H2B with free H2A/H2B dimers from nucleosomes [64, 65]. Yeast INO80-mediated genome-wide distribution of H2A.Z facilitates DNA repair, transcription, and replication [65, 66]. On the other hand, the DNA sequence-directed transcriptional activators interact with chromatin remodelers and affect their activities. For example, the interaction between yeast SWI/SNF and a DNA-bound activator can promote nucleosome eviction in vitro [67–69].

5. Aberrant chromatin dynamics and disease

To remain healthy status, our bodies must maintain homeostasis by adjusting the expression levels of genes to resist the detrimental effect from the stimuli of the biological environment both inside and outside of the body. Epigenetic modifications could regulate the expression level of corresponding gene(s), ultimately achieving homeostasis.

Chromatin dynamics play a central role in regulating various key biological phenomena. Recently, it was reported that various mutations occur in these chromatin-remodeling families, resulting in the chromatin dysregulation and aberrant expression of target genes, and ultimately lead to various disorders. Chromatin dysregulation by abnormal remodelers is often linked to neurodevelopmental disorders [70–72] and intellectual disabilities [73, 74], and also result in immunodeficiency [75] and muscle wasting syndromes [76], and various cancers [22, 77–80].

Although the pathology is beyond the scope of this chapter, to illustrate some of these possibilities, we only take the cancer, CNS disorder, and aging as examples to highlight the chromatin dysfunction mechanisms below.

5.1 Cancer

Most of the published genome-wide chromatin modification studies indicate that malignant metamorphosis of cells is governed to a large degree by the fluctuating cellular environment. Chromatin remodelers work as gatekeepers to control the accessibility of DNA binding transcription factors and ensure the variety of biological functions within the cell. Crudely speaking, the tumorigenesis can occur via at least two mechanisms: (i) altering gene expression, and (ii) fragile genome integrity and/or chromosome segregation. Thus, any aberration on chromatin remodeling complexes has been linked with genome instability. Chromosome segregation defects are related with the malignant transformation and progression of tumors. Loss-of-function SWI/SNF subunit mutations are detected in most prevalent in various cancers. It is shown that ~20% of all human cancers contain mutations on subunits of the SWI/SNF complex. The mutation of SWI/SNF subunit occur in ovarian clear cell carcinoma (75%), clear cell renal cell carcinoma (57%), hepatocellular carcinoma (40%), gastric cancer (36%), melanoma (34%), and pancreatic cancer (26%) [81, 82]. Among these ATP-dependent chromatin remodelers, SWI/SNF complex was first implicated in oncogenesis due to the discovery that its subunit SMARCB1 (also known as SNF5 and BAF47) is inactivated by biallelic mutations in nearly all cases of rhabdoid tumor [83]. Subsequently, accumulating researches reported that other subunits of SWI/SNF are mutated in various cancer types even though the mutated frequency is different in specific cancer type. For example, ARID1A is the most frequently mutated SWI/SNF subunit across cancer types [84, 85]; however, PBRM1 subunit mutations are much more common than ARID1A mutations in clear cell renal cell carcinoma [86]. Unlike the well-known role of SWI/SNF in cancers, the involvement of the other three subfamilies in cancer has not been well characterized. However, recent researches showed that all of the four chromatin remodeler subfamilies are implicated in pancreatic cancer (PDAC) by either mutation and/or chromosomal alterations [87]. Collectively, chromatin remodeling complexes is a potential target for therapeutic drugs design in the future.

Besides the point mutation or chromatin depletion, lots of mutations in epigenetic modifications occur in cancer cells compared to the normal healthy cell, which are epigenetic trademarks in earlier cancer development [88, 89]. With the abnormal epigenetic modifications, the cancer cells can maintain a portrait of self-renewal and unlimited proliferation. It has been found that cancer cells are usually marked with a loss of active H3K4me3 as well as repressive H4K20me3 and a gain of the repressive mark H3K9me3 or K3K27me3 [90, 91]. No matter altered gene expression or the instability of genome integrity during the process of tumorigenesis, questions then arise as to how do these aberrations changed the chromatin opening status and further influence the corresponding gene expression? Can we find out some methods like DNMT (DNA methyltransferases) inhibitor, HDAC (histone deacetylases) inhibitor, kinase inhibitors, etc., to reduce the detrimental effect? Or it is also worthy to modulate the balance between maturation and correction, thus favoring a status of recovery. Here, we did not expand into details and just attempt to take a short of paragraph on the chromatin dysfunction as well as some prospective pipelines for the fight against cancers.

5.2 CNS disorder

In the central nervous system (CNS) disease, epigenetic mechanisms serve as key regulators of development, homeostasis, and plasticity, all of which are highly sensitive to local and more global environmental, vascular, systemic, and intrinsic CNS factors [92, 93]. Not surprisingly, epigenetic modifications are involved in the molecular and cellular mechanisms underlying CNS pathogenesis and recovery, including the adult neurogenesis, response of initiate immunology, and neural plasticity. Disruption in the status of chromatin dynamics can lead to the changes in the site and the number of gene dysfunction. Thus, the gene regulation in chromatin level has an important role in the development of brain development. Abnormal chromatin is a key feature of necrotic cell death and apoptotic cell death, which are both associated with neural injury like stroke [94].

A growing body of evidence suggests that chromatin remodeling complexes that play a key role in vascular biology are involved in defining and transducing cardiovascular disease inheritability. The role of chromatin remodeling complexes in the transcriptional unit of protein-coding genes, especially the role of intragenic chromatin modifications, is underappreciated and not well characterized in the current era of genome-wide studies. The role of chromatin remodelers in CNS development and recovery is multifaceted. It is involved in the vulnerability of the brain cells to injury, the sensitivity of neurons to inflammation stimuli, and the immune system recovery ability after injury. Currently, epigenetic modifications mechanisms have been applied in preclinical and clinical trials due to its critical roles in the regulation of immune responses process. It also become a potential therapeutic method for risk, onset, and progression of CNS disease. The initiate immunology within brain implicated in sophisticated cognitive functions, including neuronal-glial differentiation, the modulation of neural behavior and in higher brain functions like cognition, learning, and memory. It is worth noting that epigenetic mechanisms are involved in brain immune system development, homeostasis, and plasticity.

Importantly, it has been shown the practical application of epigenetics in cardiovascular disease therapeutics [94–96]. There is increasing interest in the role of chromatin remodelers in disease pathobiology, especially about whether and how pharmacological manipulation of epigenetic processes may allow for ischemic neuroprotection [97, 98]. It is possible that epigenetic modification may serve as a sensitive and specific biomarker to predict the CNS disease progression. Furthermore, several epigenetic agents are currently being evaluated in some fields such as neural cell survival and brain tissue repair and functional reorganization. Therefore, epigenetic mechanisms have been served as key regulators for mediating neuron development, homeostasis, and plasticity.

5.3 Aging

Aging is a major risk factor for many of the most prevalent diseases all over the world [99, 100]. Epigenetic dysregulation may contribute to aging in mammals [101]. An obvious correlation between aging and DNA methylation was observed in various mouse tissues. It was reported that the genes (EDARADD, TOM1L1, and NPTX2) responsible for aging are usually hypermethylated in the promoter CpG islands [102, 103].

As we all known, at the earlier development stage of the embryonic stem cells (ESCs) occupy a global "open" and dynamic chromatin state. When the cell differentiation is on the way to be mature, the chromatin configuration will transit from "open state" to a more compact and repressive state, which correlates with less dynamic exchange of chromatin proteins [104, 105]. There are clear changes to both the global and specific histone mark patterns with organisms increasing aging. For example, H3K9me3 is a hallmark of heterochromatin, and it is globally reduced in fibroblasts from HGPS patients [106]. H4K20me3 is also a mark of heterochromatin and transcriptional repression, but it tends to increase in fibroblasts from HGPS patients with increasing aging [106]. In addition, H3K27me3 is altered in a variety of cell types and species during aging. For example, there are increased levels of H3K27me3 in brain tissue from the senescent accelerated mouse SAMP8 with increasing age [107]. A fundamental question about aging is how chromatin dynamics are passed to relatively less active through a couple of generation cell divisions.

6. Therapeutic value targeting chromatin modification

Over the past decade, rapid progress has been made in the field of epigenetics research with the development of powerful technologies such as high-resolution microscopy and genome-wide next-generation gene sequencing [108–110]. It is also very promising to use epigenetic modification changes as a diagnostic tool before the related disease develops. Several drugs designed according to the epigenetic modification have been already approved by the US Food and Drug Administration [111, 112]. Currently, epigenetic therapy is successfully applied in clinics for the treatment of hematological malignancies, but little success has been achieved in the treatment of solid tumors. However, notwithstanding the role of epigenetic regulation in the pathophysiology is not well characterized, emerging evidence suggests that it is extremely important to provide the strategies of clinic therapeutics.

7. Future perspective

Given that the various mutations occurred at different chromatin modification enzymes in different human cancers and other diseases, the investigation of specific mechanisms underlying the mutation of chromatin modification enzymes in different cancers/diseases will pave the way toward new therapeutic strategies for a range of human cancers with significant unmet medical need. In addition, from the perspective of the canonical role of chromatin remodeling complex in chromatin regulation [27], the following aspects will be the goals to develop the novel therapeutic drugs for parents with aberrant chromatin dysregulation by targeting the chromatin regulator or its associated protein. (1) designing the small molecular inhibitor against the chromatin regulators such as some specific subunits of SWI/SNF chromatin remodeling complex base on the fact that the subunits of mSWI/SNF (ARID1A, PBRM1, SMARCA4, and ARID2) are frequently mutated in many common human cancers, such as ovarian, colon, kidney, lung, prostate, breast, and others [113]; (2) designing the novel drugs targeting the transcription factor interacting with mutated chromatin remodeling complex is a potential alternative strategy to inhibit a variety of tumors driven by the interaction between oncogenic transcription factors and mutated chromatin remodeling complex; (3) given the critical role of chromatin remodeling complex in DNA replication and damage repair, it is also a potential therapeutic strategy to therapy the parents with tumor driven by aberrant chromatin regulator by targeting the replication- or repair-associated factors interacting with chromatin regulators. With increasing genetic, biochemical, and physiological understanding of chromatin remodeling complexes, their links to human diseases will continue to expand, providing new therapeutic opportunities across multiple disease areas. Given the fundamental role of chromatin regulators in normal physiology, future therapeutic approaches should focus on identifying the specific regulatory mechanisms of chromatin regulators in specific cancers/diseases to enhance overall therapeutic benefits.

8. Conclusion

Chromatin regulators are involved in priming transcriptional responses, and many chromatin modifiers and remodelers have been implicated in various human diseases. However, some chromatin alterations are potentially plastic and reversible, which raises the possibility of correcting chromatin states as a therapeutic strategy.

Probably the major debate is the association and order among gene expression, histone modification, and chromatin remodeler/dynamics, and it seemed impossible to control only one factor and hence detect the target mechanisms selectively. Further, open question reminds what is the perfect feature of these factors to guarantee a so-called healthy condition?

Despite these contentions, the progress that made by the researchers have move forward to the intrinsically epigenetic regulators with the high-throughout gene sequencing and screening technology, the scientific data supplied from various database till date will form the ladder for the future therapeutic options.

Nevertheless, studies of epigenetics are increasing, and epigenetic therapies have become exciting and promising. The rise of new technologies such as CRISPR/Cas9 gene editing and next-generation sequencing in recent years allows us to better understand the interplay among epigenetic changes, gene regulation, and human disease, and it will lead to development of new approaches for molecular diagnosis and treatments across the clinical spectrum.

The use of epigenetics as a major contributing factor in the development of normal and abnormal cells will open new sights for the advent of new therapeutic approaches.

Epigenetic therapy can be combined with the traditional therapies to provide certain treatments for reversal of the drug-resistant tumors. Also, with this therapeutic approach, the drug dosages can be reduced to eliminate the side effects of treatment and, consequently, the patient's healing problems and increase the patients' quality of life.

Author contributions

SQ had the initial idea for this article. GC and QD prepared table and drafted the manuscript. SQ and KG performed the literature search and revise support. All authors critically revised the manuscript. All authors contributed to the article and approved the submitted version.

Funding

N/A.

Conflict of interest

The authors declare that the research was conducted in the absence of any commercial or financial relationships that could be construed as a potential conflict of interest.

Abbreviation list

Snf2	Sucrose non-fermenting 2.
SWI/SNF	Switch/sucrose non-fermentable.
ISW1	Imitation switch.
CHD	Chromodomain-helicase DNA-binding protein.
INO80	Inositol requiring 80.

HAS	Helicase-SANT–associated.
RSC	Remodels the Structure of Chromatin.
NDR	Nucleosome-depleted region.
PHD	Plant homeodomain.
PWWP	Pro-Trp-Trp-Pro motif.
MBT	Malignant brain tumor.
PDAC	Pancreatic cancer.
PTMs	Histone post-translational modifications.
DNMT	DNA methyltransferases.
HDAC	Histone deacetylases.
CNS	central nervous system.
ESCs	Embryonic stem cells.

Author details

Guofei Cui[1], Qing Dong[2], Kexin Gai[3] and Shaohua Qi[4*]

1 University of Hawaii Cancer Center, Hawaii, HI, USA

2 Department of Pathology, University of California San Francisco, San Francisco, CA, USA

3 Department of Pathology, Northwestern University Feinberg School of Medicine, Chicago, IL, USA

4 Department of Neurology, McGovern Medical School, The University of Texas Health Science Center at Houston, Houston, TX, USA

*Address all correspondence to: qsh7950@gmail.com

References

[1] Strahl BD, Allis CD. The language of covalent histone modifications. Nature. 2000;**403**(6765):41-45

[2] Ridgway P, Almouzni Gv. Chromatin assembly and organization. Journal of Cell Science. 2001;**114**(15):2711-2712

[3] Mello JA, Almouzni G. The ins and outs of nucleosome assembly. Current Opinion in Genetics & Development. 2001;**11**(2):136-141

[4] Tyler JK. Chromatin assembly. Cooperation between histone chaperones and ATP-dependent nucleosome remodeling machines. European Journal of Biochemistry. 2002;**269**(9):2268-2274

[5] Kadam S, Emerson BM. Mechanisms of chromatin assembly and transcription. Current Opinion in Cell Biology. 2002;**14**(3):262-268

[6] Groth A et al. Chromatin challenges during DNA replication and repair. Cell. 2007;**128**(4):721-733

[7] Xu Y, Price BD. Chromatin dynamics and the repair of DNA double strand breaks. Cell Cycle. 2011;**10**(2):261-267

[8] Haering CH, Losada A. Understanding chromatin and chromosomes: From static views to dynamic thinking. 2013;**14**(2):109-111

[9] Nair N, Shoaib M, Sørensen CS. Chromatin dynamics in genome stability: Roles in suppressing endogenous DNA damage and facilitating DNA repair. International Journal of Molecular Sciences. 2017;**18**(7):1486

[10] Loney ER et al. Repressive and non-repressive chromatin at native telomeres in Saccharomyces cerevisiae. Epigenetics & Chromatin. 2009;**2**(1):18

[11] Papamichos-Chronakis M, Peterson CL. Chromatin and the genome integrity network. Nature Reviews. Genetics. 2013;**14**(1):62-75

[12] Lorch Y, Maier-Davis B, Kornberg RD. Mechanism of chromatin remodeling. 2010;**107**(8):3458-3462

[13] Lai WKM, Pugh BF. Understanding nucleosome dynamics and their links to gene expression and DNA replication. Nature Reviews. Molecular Cell Biology. 2017;**18**(9):548-562

[14] Mueller-Planitz F, Klinker H, Becker PB. Nucleosome sliding mechanisms: new twists in a looped history. Nature Structural & Molecular Biology. 2013;**20**(9):1026-1032

[15] Bowman GD. Mechanisms of ATP-dependent nucleosome sliding. Current Opinion in Structural Biology. 2010;**20**(1):73-81

[16] Längst G, Manelyte L. Chromatin remodelers: From function to dysfunction. Genes (Basel). 2015;**6**(2):299-324

[17] Vaquero A, Loyola A, Reinberg D. The constantly changing face of chromatin. Science of Aging Knowledge Environment. 2003;**2003**(14):Re4

[18] Kouzarides T. Chromatin modifications and their function. Cell. 2007;**128**(4):693-705

[19] Cairns BR. The logic of chromatin architecture and remodelling at promoters. Nature. 2009;**461**(7261):193-198

[20] Li B, Carey M, Workman JL. The role of chromatin during transcription. Cell. 2007;**128**(4):707-719

[21] Zhang P et al. An overview of chromatin-regulating proteins in cells. Current Protein & Peptide Science. 2016;**17**(5):401-410

[22] Clapier CR, Cairns BR. The biology of chromatin remodeling complexes. Annual Review of Biochemistry. 2009;**78**:273-304

[23] Narlikar GJ, Sundaramoorthy R, Owen-Hughes T. Mechanisms and functions of ATP-dependent chromatin-remodeling enzymes. Cell. 2013;**154**(3):490-503

[24] Hargreaves DC, Crabtree GR. ATP-dependent chromatin remodeling: Genetics, genomics and mechanisms. Cell Research. 2011;**21**(3):396-420

[25] Flaus A et al. Identification of multiple distinct Snf2 subfamilies with conserved structural motifs. Nucleic Acids Research. 2006;**34**(10):2887-2905

[26] Clapier CR et al. Mechanisms of action and regulation of ATP-dependent chromatin-remodelling complexes. Nature Reviews Molecular Cell Biology. 2017;**18**(7):407-422

[27] Reyes AA, Marcum RD, He Y. Structure and function of chromatin remodelers. Journal of Molecular Biology. 2021;**433**(14):166929

[28] Feoktistov AV, Georgieva SG, Soshnikova NV. Role of the SWI/SNF chromatin remodeling complex in regulation of inflammation gene expression. Molecular Biology. 2022;**56**(2):182-195

[29] Tang L, Nogales E, Ciferri C. Structure and function of SWI/SNF chromatin remodeling complexes and mechanistic implications for transcription. Progress in Biophysics and Molecular Biology. 2010;**102**(2-3):122-128

[30] Kassabov SR et al. SWI/SNF unwraps, slides, and rewraps the nucleosome. Molecular Cell. 2003;**11**(2):391-403

[31] Filippakopoulos P, Knapp S. The bromodomain interaction module. FEBS Letters. 2012;**586**(17):2692-2704

[32] Ichikawa Y et al. Sequence-directed nucleosome-depletion is sufficient to activate transcription from a yeast core promoter in vivo. Biochemical and Biophysical Research Communications. 2016;**476**(2):57-62

[33] Tolkunov D et al. Chromatin remodelers clear nucleosomes from intrinsically unfavorable sites to establish nucleosome-depleted regions at promoters. 2011;**22**(12):2106-2118

[34] Gai K et al. Transcriptional repression of frequency by the IEC-1-INO80 complex is required for normal Neurospora circadian clock function. PLoS Genetics. 2017;**13**(4):e1006732

[35] Yen K, Vinayachandran V, Pugh BF. SWR-C and INO80 chromatin remodelers recognize nucleosome-free regions near +1 nucleosomes. Cell. 2013;**154**(6):1246-1256

[36] Schindler U, Beckmann H, Cashmore AR. HAT3.1, a novel Arabidopsis homeodomain protein containing a conserved cysteine-rich region. The Plant Journal. 1993;**4**(1):137-150

[37] Sanchez R, Zhou MM. The PHD finger: A versatile epigenome reader. Trends in Biochemical Sciences. 2011;**36**(7):364-372

[38] Jain K et al. Characterization of the plant homeodomain (PHD) reader family for their histone tail interactions. Epigenetics & Chromatin. 2020;**13**(1):3

[39] Musselman CA, Kutateladze TG. PHD fingers: epigenetic effectors and potential drug targets. Molecular Interventions. 2009;**9**(6):314-323

[40] Lloyd JT, Glass KC. Biological function and histone recognition of family IV bromodomain-containing proteins. 2018;**233**(3):1877-1886

[41] Janzen WP et al. Epigenetics: Tools and technologies. Drug Discovery Today: Technologies. 2010;7(1):e59-e65

[42] Akhtar W et al. Chromatin position effects assayed by thousands of reporters integrated in parallel. Cell. 2013;**154**(4):914-927

[43] Carcamo CC et al. ATP binding facilitates target search of SWR1 chromatin remodeler by promoting one-dimensional diffusion on DNA. eLife. 2022;**11**:e77352

[44] Bowman GD, Poirier MG. Post-translational modifications of histones that influence nucleosome dynamics. Chemical Reviews. 2015;**115**(6):2274-2295

[45] Sawa C et al. Bromodomain factor 1 (Bdf1) is phosphorylated by protein kinase CK2. Molecular and Cellular Biology. 2004;**24**(11):4734-4742

[46] Okitsu CY, Hsieh JC, Hsieh CL. Transcriptional activity affects the H3K4me3 level and distribution in the coding region. Molecular and Cellular Biology. 2010;**30**(12):2933-2946

[47] Yan C, Boyd DD. Histone H3 acetylation and H3 K4 methylation define distinct chromatin regions permissive for transgene expression. Molecular and Cellular Biology. 2006;**26**(17):6357-6371

[48] Becker JS, Nicetto D, Zaret KS. H3K9me3-dependent heterochromatin: Barrier to cell fate changes. Trends in Genetics. 2016;**32**(1):29-41

[49] Jeziorska DM et al. DNA methylation of intragenic CpG islands depends on their transcriptional activity during differentiation and disease. 2017;**114**(36):E7526-E7535

[50] Charlop-Powers Z et al. Structural insights into selective histone H3 recognition by the human Polybromo bromodomain 2. Cell Research. 2010;**20**(5):529-538

[51] Hu Y et al. CHD3 protein recognizes and regulates methylated histone H3 lysines 4 and 27 over a subset of targets in the rice genome. Proceedings of the National Academy of Sciences of the United States of America. 2012;**109**(15):5773-5778

[52] van Nuland R et al. Nucleosomal DNA binding drives the recognition of H3K36-methylated nucleosomes by the PSIP1-PWWP domain. Epigenetics & Chromatin. 2013;**6**(1):12

[53] Musselman CA et al. Bivalent recognition of nucleosomes by the tandem PHD fingers of the CHD4 ATPase is required for CHD4-mediated repression. Proceedings of the National Academy of Sciences of the United States of America. 2012;**109**(3):787-792

[54] Mansfield RE et al. Plant homeodomain (PHD) fingers of CHD4 are histone H3-binding modules with preference for unmodified H3K4 and methylated H3K9. The Journal of Biological Chemistry. 2011;**286**(13):11779-11791

[55] Kalakonda N et al. Histone H4 lysine 20 monomethylation promotes transcriptional repression by L3MBTL1. Oncogene. 2008;**27**(31):4293-4304

[56] Boyer LA et al. Essential role for the SANT domain in the functioning of multiple chromatin remodeling enzymes. Molecular Cell. 2002;**10**(4):935-942

[57] Boyer LA, Latek RR, Peterson CL. The SANT domain: A unique histone-tail-binding module? Nature Reviews Molecular Cell Biology. 2004;**5**(2):158-163

[58] Alegría-Torres JA, Baccarelli A, Bollati V. Epigenetics and lifestyle. Epigenomics. 2011;**3**(3):267-277

[59] Du J et al. DNA methylation pathways and their crosstalk with histone methylation. Nature Reviews. Molecular Cell Biology. 2015;**16**(9):519-532

[60] Kaufman-Szymczyk A et al. The role of sulforaphane in epigenetic mechanisms. Including Interdependence between Histone Modification and DNA Methylation. 2015;**16**(12):29732-29743

[61] Swygert SG, Peterson CL. Chromatin dynamics: Interplay between remodeling enzymes and histone modifications. Biochimica et Biophysica Acta (BBA) - Gene Regulatory Mechanisms. 2014;**1839**(8):728-736

[62] Wu H, Sun YE. Epigenetic regulation of stem cell differentiation. Pediatric Research. 2006;**59**(4):21-25

[63] Voigt P, Tee WW, Reinberg D. A double take on bivalent promoters. Genes & Development. 2013;**27**(12): 1318-1338

[64] Bao Y, Shen X. SnapShot: Chromatin remodeling: INO80 and SWR1. Cell. 2011;**144**(1):158-158.e2

[65] Papamichos-Chronakis M et al. Global regulation of H2A.Z localization by the INO80 chromatin-remodeling enzyme is essential for genome integrity. Cell. 2011;**144**(2):200-213

[66] Poli J, Gasser SM, Papamichos-Chronakis M. The INO80 remodeller in transcription, replication and repair. 2017;**372**(1731):20160290

[67] Owen-Hughes T, Workman JL. Remodeling the chromatin structure of a nucleosome array by transcription factor-targeted trans-displacement of histones. The EMBO Journal. 1996;**15**(17):4702-4712

[68] Gutiérrez JL et al. Activation domains drive nucleosome eviction by SWI/SNF. The EMBO Journal. 2007;**26**(3):730-740

[69] Yudkovsky N et al. Recruitment of the SWI/SNF chromatin remodeling complex by transcriptional activators. Genes & Development. 1999;**13**(18):2369-2374

[70] Gabriele M et al. The chromatin basis of neurodevelopmental disorders: Rethinking dysfunction along the molecular and temporal axes. Progress in Neuro-Psychopharmacology & Biological Psychiatry. 2018;**84**(Pt B):306-327

[71] Lewis EM, Kroll KL. Development and disease in a dish: the epigenetics of neurodevelopmental disorders. Epigenomics. 2018;**10**(2):219-231

[72] Cotney J et al. The autism-associated chromatin modifier CHD8 regulates other autism risk genes during human neurodevelopment. Nature Communications. 2015;**6**(1):6404

[73] Kleefstra T et al. Disruption of an EHMT1-associated chromatin-modification module causes intellectual

disability. American Journal of Human Genetics. 2012;**91**(1):73-82

[74] Hoyer J et al. Haploinsufficiency of ARID1B, a member of the SWI/SNF-a chromatin-remodeling complex, is a frequent cause of intellectual disability. American Journal of Human Genetics. 2012;**90**(3):565-572

[75] Verdin E, Paras P Jr, Van Lint C. Chromatin disruption in the promoter of human immunodeficiency virus type 1 during transcriptional activation. The EMBO Journal. 1993;**12**(8):3249-3259

[76] Carr RM et al. Epigenetics of cancer-associated muscle catabolism. Epigenomics. 2017;**9**(10):1259-1265

[77] Wilson BG, Roberts CW. SWI/SNF nucleosome remodellers and cancer. Nature Reviews. Cancer. 2011;**11**(7):481-492

[78] Lai AY, Wade PA. Cancer biology and NuRD: a multifaceted chromatin remodelling complex. Nature Reviews. Cancer. 2011;**11**(8):588-596

[79] Ando M et al. Chromatin dysregulation and DNA methylation at transcription start sites associated with transcriptional repression in cancers. Nature Communications. 2019;**10**(1):2188

[80] Kukkonen K et al. Chromatin and epigenetic dysregulation of prostate cancer development. Progression, and Therapeutic Response. 2021;**13**(13):3325

[81] Shain AH, Pollack JR. The spectrum of SWI/SNF mutations, ubiquitous in human cancers. PLoS One. 2013;**8**(1):e55119

[82] Biegel JA, Busse TM, Weissman BE. SWI/SNF chromatin remodeling complexes and cancer. American Journal of Medical Genetics. Part C, Seminars in Medical Genetics. 2014;**166c**(3):350-366

[83] Versteege I et al. Truncating mutations of hSNF5/INI1 in aggressive paediatric cancer. Nature. 1998;**394**(6689):203-206

[84] Wiegand KC et al. ARID1A mutations in endometriosis-associated ovarian carcinomas. The New England Journal of Medicine. 2010;**363**(16):1532-1543

[85] Mittal P, Roberts CWM. The SWI/SNF complex in cancer - Biology, biomarkers and therapy. Nature Reviews. Clinical Oncology. 2020;**17**(7):435-448

[86] Varela I et al. Exome sequencing identifies frequent mutation of the SWI/SNF complex gene PBRM1 in renal carcinoma. Nature. 2011;**469**(7331):539-542

[87] Hasan N, Ahuja N. The emerging roles of ATP-dependent chromatin remodeling complexes in pancreatic cancer. Cancers (Basel). 2019;**11**(12)

[88] Sawalha AH et al. Defective T-cell ERK signaling induces interferon-regulated gene expression and overexpression of methylation-sensitive genes similar to lupus patients. Genes and Immunity. 2008;**9**(4):368-378

[89] Egger G et al. Epigenetics in human disease and prospects for epigenetic therapy. Nature. 2004;**429**(6990):457-463

[90] Pedersen MT, Helin K. Histone demethylases in development and disease. Trends in Cell Biology. 2010;**20**(11):662-671

[91] Messier TL et al. Histone H3 lysine 4 acetylation and methylation

dynamics define breast cancer subtypes. Oncotarget. 2016;7(5):5094-5109

[92] Szyf M. Epigenetics, a key for unlocking complex CNS disorders? Therapeutic implications. European Neuropsychopharmacology. 2015;**25**(5):682-702

[93] Knoedler JR, Shah NM. Molecular mechanisms underlying sexual differentiation of the nervous system. Current Opinion in Neurobiology. 2018;**53**:192-197

[94] Qureshi IA, Mehler MF. Emerging role of epigenetics in stroke: part 1: DNA methylation and chromatin modifications. Archives of Neurology. 2010;**67**(11):1316-1322

[95] Ordovás JM, Smith CE. Epigenetics and cardiovascular disease. Nature Reviews. Cardiology. 2010;7(9):510-519

[96] Zhang W et al. Epigenetic modifications in cardiovascular aging and diseases. 2018;**123**(7):773-786

[97] Yan MS, Marsden PA. Epigenetics in the vascular endothelium: Looking from a different perspective in the epigenomics Era. Arteriosclerosis, Thrombosis, and Vascular Biology. 2015;**35**(11):2297-2306

[98] Jhelum P et al. Implications of epigenetic mechanisms and their targets in cerebral ischemia models. Current Neuropharmacology. 2017;**15**(6):815-830

[99] Bocklandt S et al. Epigenetic predictor of age. PLoS One. 2011;**6**(6):e14821

[100] Ashapkin VV, Kutueva LI, Vanyushin BF. Aging as an epigenetic phenomenon. Current Genomics. 2017;**18**(5):385-407

[101] Krupinski J et al. DNA methylation in stroke. Update of latest advances. Computational and Structural Biotechnology Journal. 2018;**16**:1-5

[102] Jung M, Pfeifer GP. Aging and DNA methylation. BMC Biology. 2015;**13**(1):7

[103] Issa JP et al. Methylation of the oestrogen receptor CpG island links ageing and neoplasia in human colon. Nature Genetics. 1994;**7**(4):536-540

[104] López-Otín C et al. The hallmarks of aging. Cell. 2013;**153**(6):1194-1217

[105] Booth LN, Brunet A. the aging epigenome. Molecular Cell. 2016;**62**(5):728-744

[106] Shumaker DK et al. Mutant nuclear lamin A leads to progressive alterations of epigenetic control in premature aging. Proceedings of the National Academy of Sciences of the United States of America. 2006;**103**(23):8703-8708

[107] Baumgart M et al. RNA-seq of the aging brain in the short-lived fish N. furzeri - Conserved pathways and novel genes associated with neurogenesis. Aging Cell. 2014;**13**(6):965-974

[108] Xu J, Liu Y. A guide to visualizing the spatial epigenome with super-resolution microscopy. The FEBS Journal. 2019;**286**(16):3095-3109

[109] Meaburn E, Schulz R. Next generation sequencing in epigenetics: Insights and challenges. Seminars in Cell & Developmental Biology. 2012;**23**(2):192-199

[110] Lee JJ et al. Targeted next-generation sequencing reveals high frequency of mutations in epigenetic regulators across treatment-naïve patient melanomas. Clinical Epigenetics. 2015;7(1):59

[111] Heerboth S et al. Use of epigenetic drugs in disease: An overview. Genetics & Epigenetics. 2014;**6**:9-19

[112] Ganesan A et al. The timeline of epigenetic drug discovery: From reality to dreams. Clinical Epigenetics. 2019;**11**(1):174

[113] Centore RC et al. Mammalian SWI/SNF chromatin remodeling complexes: Emerging mechanisms and therapeutic strategies. Trends in Genetics. 2020;**36**(12):936-950

Chapter 2

MicroRNA Biomarkers in Primary Brain Malignancies

Denis Mustafov, Sara Seriah, Roozba Malik and Maria Braoudaki

Abstract

Despite the concerted efforts within the management of brain malignancies over the past few decades, primary brain cancers remain an obscure challenge with unfavourable outcomes for the patients. Glioblastomas (GBM) and medulloblastomas afford the most prevalent brain tumours and account for markedly high mortality rates within affected patients. The unmet clinical requirements for an early diagnostic biomarker and effective treatment have shed light onto microRNAs (miRNAs). These are small, endogenous noncoding RNAs involved in a wide spectrum of biological processes, such as post-translational modification, tumorigenesis, angiogenesis, invasiveness, and apoptosis. Increased expression of miR-21 has been shown to have devastating effects upon patients with brain tumours, and it could be used as a diagnostic biomarker and an early relapse indicator. miRNAs such as miR-128a, miR-34a, miR-7 and miR-1253 have demonstrated tumour suppressive properties and could serve as putative therapeutic agents. MiRNA signatures, such as miR-21 and miR-10b could be incorporated as potential prognostic indicators for advanced and metastatic brain malignancies, whereas miR-221/222 cluster has a therapeutic potential to sensitise cancerous cells towards radiotherapy. Herein, we summarised current knowledge on how miRNAs with significant role in glioblastomas and medulloblastomas specifically can be effectively used as promising brain cancer diagnostics, prognostics, and therapeutics.

Keywords: microRNA, signalling pathways, biomarker, diagnosis, prognosis, therapy, glioblastoma (GBM), medulloblastoma

1. Introduction

1.1 Research background

Within the current field of primary brain cancer research, a complimentary class of potential biomarkers, known as microRNAs (miRNAs), are becoming increasingly favoured upon their pleiotropic advantages; from adopting potential diagnostic, prognostic, and therapeutic properties. In the contemporary study of the adult brain tumour, glioblastoma (GBM), several miRNAs are seen to act as potential biomarkers of the debilitating cancer. Through the revelation of current studies, miR-21, miR10b, miR-221, and miR-222 are seen to have promising usages of becoming novel diagnostic and prognostic markers. Besides the use of individual miRNAs to act as biomarkers, the cluster, miR-221/222, has also been seen to additionally possess therapeutic potentials within the treatment of high-grade gliomas. Similarly, miRNAs can also serve as biomarkers within medulloblastomas, a paediatric brain cancer. The miRNA, miR-10b, is seen to own diagnostic and therapeutic potentials within a subgroup of medulloblastoma, known as Sonic Hedgehog (SHH). Additionally, miR-466-3p can act both therapeutically and diagnostically, dependent on the regulation pattern of the miRNA. The review will expand upon the potential usages of the miRNAs mentioned above but will also introduce additional potential miRNAs to highlight the beneficial impacts this class of noncoding RNAs can play within primary brain malignancies.

2. Primary brain malignancies

Classed as a heterogenous set of tumours, primary brain cancers are termed as abnormal cellular growths within the cavity of the brain [1]. Although the malignancy is classified as rare, primary brain cancers owe to significantly high mortality and poor survival rates, with only 40% of patients surviving over a year [2]. The incidence rates for brain cancers in the UK alone has increased by 39% since the 1990s [3]. In fact, by 2035, it is estimated that incidence rates will increase by a further 5% and 8% for males and females in the UK from 2014 [3]. Similarly, individuals affected with brain cancers are more likely to suffer from prolonged life-changing cognitive, physical and psychological impairments unlike other types of cancers, with one study observing approximately 90% of patients with brain metastases displaying substantial cognitive deficits prior to treatment [4, 5]. Thus, there is a demand for novel interventions to deliver successful treatment and improve prognosis outcomes for patients.

Brain malignancies in adults currently stand as the eighth most common cancer [4]. In fact, the tenth leading source of deaths for men and women is from brain and central nervous system (CNS) cancers [6]. Serving as a highly heterogeneous tumour, GBM is a rare aggressive adult primary brain cancer [7]. Native to gliomas, the malignancy subtype collectively constitutes 81% of malignant intracranial tumours in adults [4]. Despite significant efforts, the five-year survival rates for GBM patients remain low standing at only 6.8%, with an 8-month median survival period on average [4]. Often a full brain tumour resection is not achievable due to the anatomical structure of the cranial and brainstem nerves surrounding the brain [8]. Even with surgical interventions to increase prognosis outcomes, the majority of GBM tumours remain obstinate to chemotherapeutics; a primary cause for the reduced efficacy outcomes for GBM patients [9].

Among children to adolescent years, paediatric brain cancers stand as the leading source of cancer-related deaths [4]. Categorised as solid tumours, risk factors to the young are thought to be influenced from environmental and genetic factors; where a family history and a maternal age over 40 during birth, as well as high radiation exposure commonly seen in leukaemia patients, all pose high risks to the child [10]. Medulloblastoma, a high-grade tumour, accounts for approximately 10% of all paediatric brain malignancies [11]. Associated with significantly high morbidity rates, the malignant tumour originates within the posterior fossa of the brain [11]. Common treatment plans for medulloblastoma patients include chemotherapy and/or radiotherapy. Although, a study observed the nutritional effects young children gained from chemotherapy, where the nutritional status of patients began to fall; inclining them to a mean weight loss of 8.2% during the second course of chemotherapy treatment from diagnosis [12]. Additionally, surgical interventions are also primarily used to exile tumours; however, it has been reported that up to 40% of patients suffer from neurological losses from surgery [13]. Even with successful treatment outcomes, both children and adults can endure long-term debilitating and neurological effects. Paediatrics can suffer from learning and growth difficulties, while adults tend to be stranded with cognitive and neurological impairments [13]. Since current findings highlight the significant association between morbidity and current treatments for medulloblastoma patients, there is a need for novel interventions which allow for long-term effective treatment plans for patients.

In the management of brain malignancies, the location of primary brain tumours has shown to pose mainstream challenges to current therapeutic interventions, restricting efficacy outcomes in patients. Majority of current drug and chemotherapeutic drug delivery strategies have been shown to have difficulty in the passive movement across the blood brain barrier (BBB), due to intracranial endothelial cells forming tight junctions which limits the passive diffusion to only small sized gas and lipophilic molecules [14]. In turn, this poses constraints for chemotherapy treatments and the readily delivery of therapeutic drugs to brain tumours, thus restricting treatment efficacy. In order to overcome the microvasculature structure, current studies have begun to observe the successful usages of nanoparticles, intra-arterial and intranasal methods of treatment deliveries through the BBB [15]. However, further research is required to optimise the scope of new techniques for drug deliveries to patients to improve current survival rates post treatment.

3. MicroRNA biogenesis

MiRNA biogenesis begins with post-or co- transcription of transcripts of RNA polymerase II/III [16]. At present, the majority of miRNA identified are intragenic and commonly processed from noncoding regions of protein coding genes, known as introns [17]. The remaining miRNAs are referred to as intergenic; where both, transcription occurs independently from a host gene and regulation occurs from their own independent promoters [17]. From the study of miRNAs, the pleiotropic nature of these noncoding RNAs, which have significant utilities within disease states and drug resistance, elucidates the potential of these molecules to serve as important biomarkers within the diagnosis and treatment of a wide range of diseases [18].

The classification of miRNA biogenesis is separated into two distinct pathways: non-canonical and canonical. The primary route for miRNA processing occurs in the canonical biogenesis pathway. The recognition of multiple motifs and N6-methyladenylated GGAC occurs primarily from the DiGeorge Syndrome Critical Region 8 (DGCR8) [19]. The protein alongside Drosha, a ribonuclease III enzyme, forms a microprocessor complex, allowing the processing of transcribed primary miRNA (pri-miRNA) into precursor miRNA (pre-miRNA) [20]. The resulting catalytic subunit of the microprocessor complex, Drosha, can cleave pri-miRNA from its hairpin assembly, forming 2-nucleotide 3′ overhang on the pre-miRNA [21]. The overhang of pre-miRNA exports from the nucleus into the cytoplasm with the assistance of exportin 5 (XPO5)/RanGTP complex [22]. The pre-miRNA becomes processed by Dicer, a RNase III endonuclease, where a mature miRNA duplex is biosynthesised through the elimination of the terminal loop [23]. Depending on whether the guide strand arises from the 3′ or 5′ end which is reliant on the either use of 3p or 5p strand, the Argonaute protein (AGO 1–4) facilitates the loading of both strands onto the protein [24]. This results in the formation of an additional complex known as the miRNA-induced silencing complex (RISC), where miRNAs have the ability to bind to 3′ untranslated regions (UTR) and thus, become regulators at a post-transcriptional level [17].

Conversely, the non-canonical pathway utilises predominant proteins from the canonical biogenesis pathway, such as Dicer, AGO2, Drosha and exportin 5 [16]. However, the pathway can be further sub-categorised into Drosha/DGCR8-independent and Dicer-independent routes [25]. Among the Drosha/DGCR8 independent pathway, pre-miRNAs such as mirtrons feature substrates of the RNase III endonuclease, Dicer to complete cytoplasmic maturation [26]. Unlike the canonical pathway, Drosha cleavage is not utilised and instead the nascent RNAs leave the nucleus to the cytoplasm via exportin 1 [16]. Similarly, the Dicer-independent pathway processes miRNA from transcripts of endogenous short hairpin RNA (shRNA) via the ribonuclease III enzyme, Drosha [27]. Within the cytoplasm, AGO2 allows maturation of the pre-miRNA, since they cannot resemble the sufficient length of Dicer substrates [27]. The maturation process ends when the 3p strand is sliced via AGO2 and the 5p strand becomes trimmed via 3′ to 5′ trimming [28].

Following the production of minimal miRNA-induced silencing complex (miRISC) in the cytoplasm, miRNA response elements (MREs) allow miRISC to maintain target specificity [29]. The 3′ untranslated region is typically where a regulatory potential is provoked, though MREs are situated throughout a mRNA molecule [30]. The degree in which a given target can be controlled is crucially dependent on miRNA and MRE affinity for one another, subcellular location and number of miRNA and MRE present in the cell [30]. However, the vast number of miRNA:MRE interactions within animal cells are not entirely correlative to each other. Since, MREs commonly have at least one central mismatch on the guide miRNA, the function of AGO2 endonuclease is inhibited [16].

4. MicroRNAs in brain malignancies

MiRNAs are a novel class of endogenous noncoding molecules of RNA composed of a nucleotide length of 18–22 [7]. These single stranded molecules facilitate the expression of target genes at a post transcriptional level through complementary base pairing with specific regions in target mRNAs [9]. Initially discovered in the nematode *Caenorhabditis elegans*, miRNAs are becoming recognised, with an upward interest in miRNA novel therapies in cancers, upon their diagnostic and therapeutic potentials. The expression of various functioning genes can become downregulated by a single miRNA. Hence, the development of novel methods to potentially identify and alter miRNA pathways can lead to a new discovery in cancer treatment, since cancers contain numerous gene aberrations [31]. In contrast to conventional drug therapies, miRNAs can pass more easily through the BBB into the intracranial space, particularly in disease states [32]. Thus, miRNAs can become optimally used as potential biomarkers, as well as in novel drug delivery mechanisms into intracranial tumours. **Figure 1** below outlines the scope of this chapter demonstrating the possible uses of miRNAs in combating brain malignancies.

With poor long-term survival outcomes for primary brain cancers, the standard treatments of surgery and radiotherapy remain unsatisfactory. However, miRNA studies provide potential hopes from numerous novel approaches in the treatment of primary brain malignancies. Firstly, current miRNA studies highlighted the potential

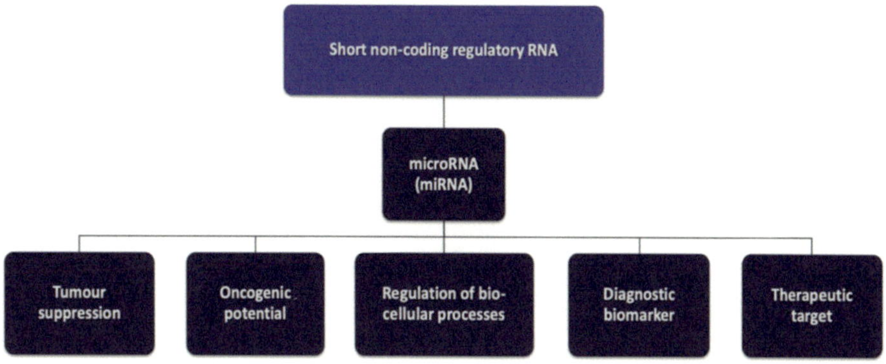

Figure 1.
Scope of the possible implications of miRNAs in combating brain malignancies. These small, non-coding RNAs have shown the ability to act as oncomiRs or tumour suppressors. Thus, their elevated or decreased levels could be possibly used as diagnostic, prognostic, and therapeutic markers in the management of CNS tumours.

ability of a miRNA to class tumours, with a study by Lu et al., finding expression profiles of miRNA to be effective in classifying poorly differentiated cancers [33]. The miRNA profiles exhibited greater levels of knowledge regarding tumour states and lineage progression of tumours [33]. Thus, from the successful usages of miRNAs in the differentiation of cancers, these pleiotropic molecules can potentially become used within the development of future novel therapeutics in cancer.

Additionally, miRNAs are increasingly becoming known to exhibit tumour suppression properties within brain cancers, a potential approach for treatment. A recent study by Xue et al. found that the tumour development from medulloblastoma was downregulated by two exosomal miRNAs; miR-101-3p and miR-423-5p. These tumour suppressors targeted the *FOXP4* gene and the histone methyltransferase, *EZH2* [34]. Furthermore, a contemporary study revealed the therapeutic potential usages of the tumour suppressor miR-138, for treating primary glioblastoma, by directly decreasing the regulation of CD44 to suppress proliferation [35]. Moreover, a study carried out by Costa et al. observed the potential advantages of anti-miR-21 oligonucleotide within a glioblastoma mouse model. The miRNA was found to decrease cellular proliferation and tumour growth, while being able to also increase apoptosis within the model [36].

Therein, miRNAs hold therapeutic capabilities to differentiate tumour grades, excel the current standards of knowledge regarding these cancers, as well as acting as tumour suppressors in the treatment of primary brain tumours. Such novelties should be explored in the treatment of such debilitating tumours, to provide personalised therapies for many primary brain malignancies. Although additional research is required to demonstrate miRNA therapeutic potentials in clinical context, the discovery of miRNAs to possess multiple advantageous properties promise a forward approach towards the treatment and outcomes for primary brain cancers.

5. MiRNAs and glioblastomas

5.1 miRNAs as biomarkers within glioblastomas

The unmet clinical requirements for an early diagnostic tool and effective treatment for glioblastoma via the existing routine strategies have initiated the need of novel approaches for the early and correct diagnosis of GBM, followed by an adequate prognostic plan and a possible treatment strategy. The extensive role of miRNAs in the regulation of GBM tumorigenesis has made these small, non-coding RNAs an attractive source of information for researchers. MiRNAs found in GBM are involved in a wide spectrum of biological processes ranging from neurone differentiation and maturation, post-translational modification of genetic information, tumorigenesis, angiogenesis, invasiveness, resistance to treatment, apoptosis, and immune system modulation. In a systematic review Møller and colleagues demonstrated that more than 300 miRNAs are deregulated in GBM, with miR-253 being overexpressed and miR-95 under-expressed [37]. Faulty events during the biogenesis of miRNAs can lead to their deregulation in many cancers. Such events include amplifications, deletions, epigenetic modifications, translocations, and silencing of miRNAs [38].

5.2 Overexpressed miRNAs in GBM and their possible implications as biomarkers

Some miRNAs can act as tumour suppressors and others as oncogenes (oncomiRs). Detection of miRNA signatures in primary brain tumours has revealed unique avenues for assessing the diagnosis, prognosis, and monitoring of patients [39]. Some of the well-studied examples of overexpressed miRNAs in high grade glioblastomas with diagnostic and prognostic properties include miRNA-21, miRNA-10b, miR-221, and miR-222. The molecular mechanisms via which these miRNAs act are under extensive evaluation with some of the affected genes and pathways shown in **Figure 2**.

miR-21 is aberrantly expressed in many types of cancer, such as colorectal, lung, pancreas, leukaemia and GBM. Located within the vacuole membrane protein 1 (VMP1) locus on chromosome 17, the mature miR-21 is transcribed in a complex

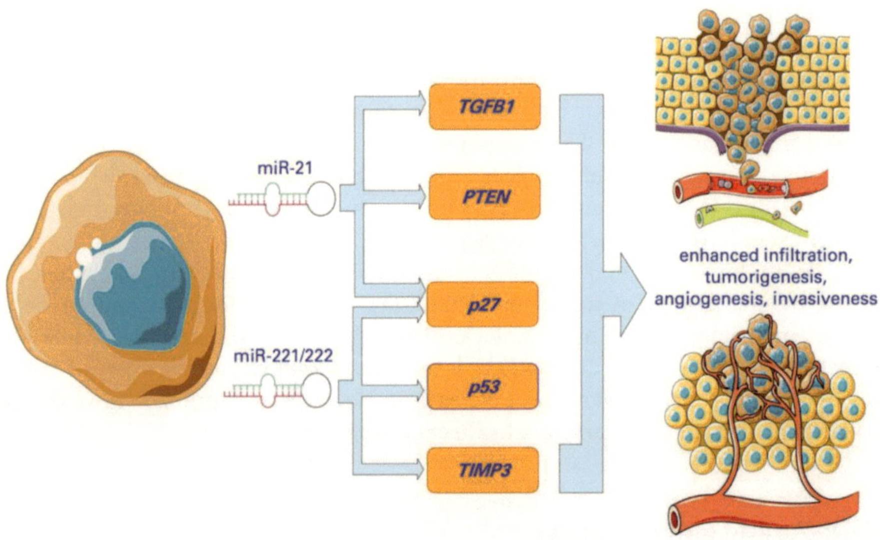

Figure 2.
Molecular pathways affected by miR-21 and miR-221/222. Overexpression of oncomiR-21 leads to enhanced cell survival, and thus proliferation, due to the inhibiting ability of miR-21 towards pro-apoptotic genes such as PTEN. miR-221/222 directly regulate glioma cells invasion via the tissue inhibitor of metalloproteinase 3 (TIMP3). miR-221/222 lower the gene expression and diminish the protein levels of p27 and 57, thus promoting S-phase progression and cell proliferation.

manner from two pri-miR-21 of an approximate size 3.5 kb and 4.3 kb [40]. miR-21 plays a pivotal role in the tumorigenesis of GBM and is the only miRNA, which to date has clearly defined diagnostic and prognostic properties. The levels of miR-21 drastically decrease post tumour resection, allowing the biomolecule to be used as monitoring agent for patients and detect early relapse [41]. High levels of miR-21 in blood samples predict a poor prognosis for GBM patients, due to the pro-tumorigenic properties of miR-21, acting as a down-regulator for insulin-like growth factor-binding protein 3 (IGFBP3) and other caspases, thus inhibiting programmed cell death [42]. Multiple studies have demonstrated the up-regulated profile of miR-21 in several types of cancers, indicating its potential role as an oncomiR [43, 44]. Extracellular miR-21 has demonstrated its relevance as a diagnostic biomarker of GBM. A meta-analysis study has revealed that miR-21 can predict GBM with high accuracy and specificity [45]. The unique properties of miR-21 in gliomas differentiated between other brain tumours and have been validated by Ivo D'Urso and colleagues who showed that miR-21, alongside miR-16 possessed a 90% sensitivity and 100% specificity in doing so [46]. The detection of miR-21 in cerebrospinal fluid (CSF) and its diagnostic relevance has also been evaluated via its direct action upon the TGF-β/Smad3 signalling pathway. The researchers found out that the inhibitor for GF-β type I receptor kinase, named galunisertib, decreased the expression of miR-21 and thus supressed its oncogenic properties [47]. However, screening CSF samples is not an easy task and might lead to complications, if not performed precisely. Elevated levels of miR-21 were also reported following examination of the plasma samples collected from GBM patients in which the results showed increased expression of miR-21 [48]. Mao and colleagues also studied the expression of miR-21 in serum samples from control and GBM patients and demonstrated that miR-21 was significantly upregulated in all GBM samples [49]. The overexpression of miR-21 in glioblastoma tumours has shown to lead to enhanced cell survival, and thus proliferation, due to the inhibiting ability of miR-21 towards pro-apoptotic genes such as *PTEN*, consequently leading to avoiding apoptosis [44, 50].

Another miRNA with a high oncogenic potential and universal to GMB is miR-10b. miR-10b is virtually undetectable in normal brain tissue; however, in low- and high-grade brain tumours from different subtypes this miRNA becomes abundantly expressed [51]. Located within the *HOXD* genomic locus, miR-10b is involved in various cancerogenic pathways including proliferation, invasion, and metastasis of malignant glioblastomas. Although miR-10b has been found to be deregulated in different types of tumours such as ovarian and gastric tumours, alongside glioblastomas, the regulation of the miRNA appears to be cell- and context-specific [52]. The detection of high levels miR-10b in the serum of GBM patients who have undergone a treatment therapy with bevacizumab has indicated the potential of miR-10b as a prognostic marker for monitoring therapy. The researchers identified that miR-10b, alongside miR-21, were highly expressed in GBM patients' post-treatment in comparison to pre-treatment levels of both miRNAs [53]. A negative correlation between the highly expressed miR-10b and miR-21 and the size of glioblastomas was also evaluated throughout this research. The same correlation was not observed in patients who have undergone a temozolomide therapy. Thus, miR-10b, in a combination with miR-21, might be incorporated in monitoring patients treated with bevacizumab. While miR-10b is specifically detectable in the CSF of patients with advanced and metastatic brain tumours, miR-21 is expressed in various cancer types and normal brain tissue and lacks exclusive specificity for GBM [54]. Low or absent levels of miR-10b and miR-21 were found in the CSF of GBM patients in remission, with an increase of both miRNAs during relapse rates and progression of the tumour with an accuracy 91–99% [51]. Thus, the utilisation of miR-10b alongside miR-21 could be a possible monitoring and prognostic signature biomarker of advanced and metastatic GBM.

The overexpressed miR-221/222 cluster is associated with the degree of glioblastoma infiltration and poorer overall survival [55]. The collectively encoded miR-221 and miR-222 in a gene cluster located on chromosome X (Xp11.3) are highly conserved in vertebrates with an identical seed region separated by 727 bases [56]. The role of the miRNA cluster as prognostic marker in glioblastomas has been demonstrated by Zhang and colleagues, who identified significantly high plasma levels ($p = 0.0001$) of miR-221/222 in glioma patients which positively correlated with poorer survival rates within 95% of the studied cohort [57]. Some of the molecular mechanisms via which the miR-221/222 family acts during glioblastoma carcinogenesis are by promoting the S-phase of the cell cycle, inhibiting apoptosis, or regulating the invasiveness of the cancer. The up-regulation of miR-221/222 is closely related with the cell cycle check points *p27* and *p57*. Both miRNAs bind to their 3' UTR regions ensuring lower gene expression and diminish the protein levels of p27 and 57. This in turn promotes S-phase progression and cell proliferation [58]. miR-221/222 were shown to directly regulate glioma cells invasion via the tissue inhibitor of metalloproteinase 3 (TIMP3). The researchers demonstrated that TIMP3 is a direct target for miR-221/222 and knockdown of miR-221/222 in xenograft mouse models restored the normal levels of TIMP3 and reduced tumour growth [55]. Researchers also demonstrated that the cluster could potentially serve as a therapeutic agent by increasing the radiosensitivity within glioblastoma cells via *PTEN* independent activation of the Akt pathway [59]. Tokudome and colleagues have identified low levels of *PTEN* after radiotherapy, suggesting the miR-221/222 cluster could act as an inhibitor for glioblastoma cells post radiation and thus supress tumour growth [60]. The significance of the prognostic and potential therapeutic implications of miR-221/222 in high-graded gliomas is supported by their interactions with the tumorigenic genes *TIMP3*, *p27*, *p53* and *PTEN*. The plasma levels of miR-222 and miR-21 were shown to be reduced after total tumour resection within GBM patients. Thus, a signature of miR-222, miR-21 and miR-124-3p could find an implication in monitoring patients with post tumoral resection and possibly identify early relapse [61].

5.3 Under-expressed miRNAs in GBM and their possible implications as biomarkers

MiRNAs with potential diagnostic properties include miR-128, miR-34-3p, and miR-7 [45]. These miRNAs are downregulated in glioblastoma patients and act as tumour suppressive agents that could be used for diagnostic, prognostic and therapeutic purposes. The molecular mechanisms via which these miRNAs regulate some of the affected genes and pathways shown in **Figure 3**.

The brain-enriched miR-128 is a type of an intronic miRNA encoded by two different genes, miR-128-1 and miR-128-2, located on chromosomes 2q21.3 and 3p23.p, respectively [62]. The normal expression miR-128 has been linked to normal brain development [63]. However, miRNA assays, quantitative Real-Time Polymerase Chain Reaction (RT-qPCR) and Western blot analyses indicated that miR-128 was under-expressed in aggressive solid brain tumours, including glioblastomas and medulloblastomas, when compared to normal adjacent brain tissue [64]. miR-128 exerts its role in glioblastoma tumorigenesis via different pathways, including inhibition of proliferation, influencing apoptosis and drug resistance, regulating epithelial to mesenchymal transition and inhibiting tumour cell invasion and motility. Previous

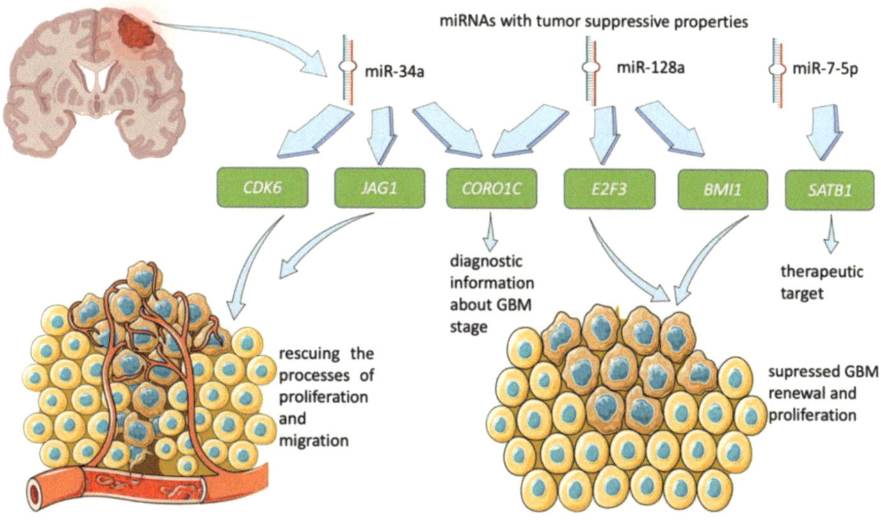

Figure 3.
Molecular pathways affected by miR-128a and miR-34a. miR-128a is associated with the polycomb complex protein BMI1 and E2F transcription factor 3, E2F3. miR-128a can inhibit proliferation of glioblastoma cells by directly targeting E2F3a. miR-128a could supress GBM renewal and proliferation by directly targeting BMI1. The relevance of CORO1C in glioblastoma has demonstrated a link to the grade of the malignancy. The consequences of the downregulation of miR-34a in GBM has been associated with the suppression of several oncogenes, including CDK6, Notch1 and Notch2. The induced overexpression of miR-34a possess the potential to supress the functions of distorted genes, such as JAG1 and CDK6, rescuing the processes of proliferation, cell cycle progression, survival, and migration.

research has confirmed the high analytical specificity and sensitivity of both, miR-21 and miR-128, in their use as diagnostic markers for GBM. Roth and colleagues demonstrated the significant downregulation of miR-128 in the peripheral blood samples of 20 GBM patients with an accuracy of 81%, and sensitivity and specificity of 79% and 81%, respectively [65]. However, the number of studied patients was only 20, indicating the need of validating these results in a larger cohort of GBM patients. The forced expression of miR-128a demonstrated inhibition of GBM development via the promotion of apoptosis in the U-87 MG GBM cell line [66]. However, the molecular mechanism via which miR-128a acts in glioblastoma multiform tumours is not yet fully understood. Multiple studies have suggested that the expression of miR-128a is differentiated in various types of cancer and might exert distinctive roles in cancer development. For instance, miR-128a has been found to be overexpressed in acute lymphoblastic leukaemia [65, 67]. miR-128a was mainly found to be associated with the polycomb complex protein BMI1 and E2F transcription factor 3, *E2F3*. Previous experiments have demonstrated that miR-128a can inhibit proliferation of glioblastoma cells by directly targeting *E2F3a* which could subsequently lead to rescuing the supressed proliferation mechanism found in GBM [68]. The link between BMI1, a stem cell renewal factor and miR-128a has also been documented by Godlewski and colleagues, who demonstrated that miR-128a could supress GBM renewal and proliferation by directly targeting BMI1 [69]. The relevance of coronin-1 C, (*CORO1C*), in glioblastoma has demonstrated a link to the grade of the malignancy [70]. Cell cycle progression, cell transduction and apoptosis are possible pathways via which the gene might exert its tumorigenic potential. The expression of the gene has been examined in different types of brain malignancies. Hence, CORO1C might provide useful diagnostic information about GBM stage and could possibly be blocked when miR-128a is overexpressed, and thus modify its oncogenic activity.

The tumour suppressor miR-34a has emerged as a possible therapeutic agent for GBM patients. Located on the second exome of chromosome 1p36, miR-34a is encoded by its own, two transcripts, which are highly conserved in humans [71]. This miRNA is thought to act as a tumour suppressor via the p53 pathway as its gene promoters contain *p53* binding sites. Researchers identified significantly lower levels of miR-34a in patients with mutated *p53* status in comparison to wild-type *p53* GBM samples [72]. Gao and his colleagues also identified that the levels of miR-34a in high grade gliomas were significantly lower when compared to normal adjacent brain tissue. This finding allowed for the consideration of miR-34a as a possible diagnostic and predictive biomarker in GBM. The under expression of miR-34a has been shown to correlate with poorer prognosis in GBM patients. The consequences of the down-regulation of miR-34a in GBM has been associated with the suppression of several oncogenes, including *CDK6, Notch1* and *Notch2* [71]. Researchers have suggested that *Notch1* and *Notch2* are frequently overexpressed in glioblastomas and medulloblastomas [73]. As a key player in cell-to-cell communication, normal neuronal development and differentiation, and *de novo* blood vessel formation, the dysregulated Notch signalling pathway is a very important tumorigenic factor in GBM. For the Notch signalling pathway to transduce signals between cells, a family of Jagged protein receptors embedded in the membranes of adjacent cells are required. One of these protein receptors is Jagged1, transcribed from the *JAG1* gene. The oncogene cyclin dependant kinase 6, *CDK6*, a serine/threonine protein kinase that regulates transition through the cell cycle has been shown to be overexpressed in brain tumours. The expression of *CDK6* at later tumour stages was found to be increased in 12 out of 14 glioblastoma tumour samples [74]. Increased levels of miR-34a *in vitro* could induce apoptosis and inhibit proliferation in GBM cell lines [75]. Thus, the induced over-expression of miR-34a possess the potential of a therapeutic miRNA that could be used to supress the functions of distorted genes, such as *JAG1* and *CDK6*, rescuing the processes of proliferation, cell cycle progression, survival, and migration. The lack of research investigating these genes indicates a gap in this field of cancer research, with the potential for discovering new molecular mechanism via GBM tumorigenesis affecting possible targeted therapeutics.

Another miRNA that has been found to be downregulated in glioblastoma tissues is miR-7. In humans, an identical mature sequence of this miRNA can be encoded by, miR-7a-1, miR-7a-2, and miR-7b, located on different chromosomes [76]. miR-7 is highly expressed in normal brain tissue and plays an important role in many physiological and pathological processes within the brain. miR-7-5p was found to inhibit cell migration and invasion in glioblastomas by targeting the special AT rich sequence binding protein (SATB1) [77]. Yin and colleagues demonstrated that miR-7-5p has a suppressive effect upon SATB1 within the U87 and U373 glioblastoma cell lines expressed in inhibited migration and invasion of the cells. An immunohistochemical analysis of a microarray with 122 glioma samples has indicated that high-grade gliomas were associated with significant expression of phosphorylated SATB1, which in turn also correlated with poorer overall survival rates indicated by Kaplan-Maier analysis [78]. The suppressive effect of miR-7-5p upon SATB1 provides a potential avenue for treatment. Delivery of miR-7-5p in DNA-cationic liposome complexes to glioblastoma cells demonstrated significant growth and metastasis inhibition *in vivo* [79]. The direct inhibiting action of this tumour suppressive miRNA upon EGFR antagonises downstream effectors such as *ERK*, *Akt* and *Stat3* which subsequently leads to enhanced apoptosis and ceased inhibition, proliferation, and migration within glioblastoma cells. Kefas and colleagues demonstrated low levels of miR-7 in glioblastoma tissues and further evaluated its inhibiting action towards the EGF receptor subsequently leading to impaired viability and metastatic properties of GBM cells [80]. miR-7-5p has the potential to be incorporated into new, targeted therapies within patients expressing wild type EGFR glioblastoma molecular profiles.

Despite the efforts made to combat the deadly glioblastoma multiform tumours, this type of brain cancer is still associated with poor prognosis and low overall survival rates. MiRNAs have demonstrated promising outcomes in their use as prognostic, diagnostic biomarkers and treatment targets. However, their clinical adaptation is still far from accepted, due to uncertainties in the experimental findings caused by the limited number of patient samples used in research [81].

6. MiRNAs in medulloblastomas

6.1 MicroRNAs as potential biomarkers in paediatric medulloblastomas

Similar to several types of paediatric brain cancers, miRNAs are involved in the regulation of different cellular and physiological processes in medulloblastoma including CNS development related processes. Over 60% of the reported miRNAs are detected in the adults' brain and their expression changes as the brain goes through maturation and develops from embryonic to adult stages [82]. They are involved in the regulation of the post translational process that controls the neural development and morphology. Thus, they play a pivotal role in cellular events related to promoting or supressing tumour growth and proliferation either as oncogenes or tumour suppressors [82].

For instance, miR-124 is reported to be one of the most expressed miRNAs in the mature CNS. It also plays a crucial role in the neural differentiation and maturation [83].

MiR-124 was also reported to have an important role in normal prefrontal cortex (PFC) and brain functioning as it regulates the Dopamine D2 receptor (Drd2) pathway which is responsible for dopamine regulation and secretion. Studies have reported a relation between decreased mi-124 expression and brain disorders including Alzheimer's disease and frontotemporal dementia (FTD) [83, 84]. In mice exposed to chronic ultra-mild stress, an overall of 80% decrease -compared to non-stress exposed mice- in mir-124 expression was observed and depression-like behaviours were exhibited [84–86].

Despite the unclear mechanism of action of miR-124 in normal brain and based on the previously reported findings and studies, miR-124 can afford to be a potential diagnostic biomarker and therapeutic target for CNS disorders and brain cancer [87].

6.2 MiRNAs in different medulloblastoma subgroups

As the most common severe paediatric brain malignancy, medulloblastoma has four molecular subtypes: Wingless (WNT), Sonic Hedgehog (SHH), Group 3 (Gr3) and Group 4 (Gr4). Based on the aforementioned discussion, miRNAs play a crucial role in the neuron development and maturation. Hence, they have a share in promoting or supressing tumour growth by their aberrant expression. However, an entire clear and detailed role description of miRNAs in tumours remains to date unclear [82].

6.2.1 Wingless (WNT) subtype

Several miRNAs were reported to be downregulated in WNT subtype including miR-383, miR-206, miR-183, miR-128a/b, miR-449, and miR-133b. Tumour formation initiated by miRNA downregulation indicated that they act as tumour suppressors [82, 88]. Thus, miRNAs with tumour suppressive effect could afford potential diagnostic biomarkers and promising therapeutic targets by upregulating their expression and restoring their tumour-suppressing function. For instance, miR-148a expression was reported to reduce Neuropilin (NRP1) expression that is involved in several pathways promoting tumour growth and metastasis. Considering its suppressive effect on tumour promoting pathways and factors including NRP1, miR-148a is considered one of the main reasons behind the lower metastatic incidence and good survival rates of the WNT subtype patients. The downregulation of the NRP1 by the miR-148a suggested a good diagnostic biomarker and a highly promising therapeutic agent for this medulloblastoma subtype [88–90].

6.2.2 Sonic hedgehog (SHH) subtype

The SHH subtype has a moderate prognosis that depends on the molecular mutation and the metastatic status. Alteration in the SHH signalling pathway results in tumour formation, development, and proliferation [82]. Among these mutations, protein patched homologue (PTCH) inactivating and smoothened homologue (SMO) activating mutations are the two most common mutations. Patients diagnosed with SHH medulloblastoma and have additionally a *TP53* gene mutation have the worse outcome. Around 80% of SHH cases combined have mutations in the downstream SMO pathway, resulting in tumours that are resistant to SMO inhibitors. Also, the Nrp2 receptor and its ligand Vegfa are up regulated in SHH's cancer stem cells (CSCs) promoting their self-renewal ability and viability [10]. Among the validated inhibiting molecules of Nrp2 and Vegfa molecules is the miR-446-3p. Stated that, an upregulated expression of miR-466-3p could potentially be considered as a therapeutic candidate while its downregulated expression could afford being a diagnostic biomarker [82, 91].

With its exclusive expression in tumours, the miR-10b is another miRNA that plays a crucial role in medulloblastoma cell proliferation, invasion, and survival by controlling B cell lymphoma 2 (*BCL2*) levels. The *BCL2* regulates apoptosis and is maintained in balanced levels in healthy cells. The miR-10b oncomiR affects the modulated apoptotic function of *BCL2* and promotes cancer cell survival. Thus, miR-10b could serve both as a good diagnostic biomarker and therapeutic target for SHH medulloblastoma subgroup [92].

6.2.3 Group 3 (Gr3) and group 4 (gr 4) subtypes

On one hand, the most aggressive and yet the least understood subtype of medulloblastoma is group 3 MB. Around 45% of group 3 MB cases are metastatic at diagnosis stage and most cases are resistant to adjuvant therapies which results in poor prognosis, and low survival rates. On the other hand, group 4 medulloblastoma has better survival rates, also known as intermediate and better prognosis than group 3 MB, regardless of the 40% of cases that are identified/classified as metastatic at diagnosis [93, 94].

In contrast to WNT and SHH subgroups, both group 3 and 4 MB have no distinguishing altered signalling pathways and no signs on known molecular mutation

MB subtype	miRNA	Expression	Targeted genes
All subtypes	miR-21	Upregulated	*PDCD4*
All subtypes	miR-106b		*PTEN*
SHH	miR-183-96-182 cluster		*SHH*
SHH and Group 3	miR-10b		*BCL2*
WNT	miR-224		*WNT*
WNT	miR-193		*WNT*
SHH	miR-124	Downregulated	*SLC16A1*
SHH	miR-324		*SHH*
Group 3/4 SHH	miR-192		*DHFR, CD47*
SHH	miR-128a		*BMI-1*

Table 1.
Different miRNAs expressed in the four subtypes of medulloblastoma [82, 83, 93–100].

including TP53 mutations are observed in both groups. Therefore, challenges regarding diagnosis and therapeutic targets sets on continuous research to identify the mechanisms of origination and development of these MB subgroups [82, 93, 94]. A very low number of expressed miRNAs associated with MB groups 3 and 4 have been identified. For instance, miR-1253 is a brain-enriched microRNA that plays a key role in regulating bone morphogenic proteins during cerebellar development. The increased expression of the miR-1253 has been associated with the activation of apoptotic pathway and reduction of tumour malignancy. The study that stated the aforementioned tumour suppressive properties of miR-1253 has also stated that it is a good potential diagnostic biomarker for mainly groups 3 and 4 by its low expression in tumour cells. The study has also mentioned the promising therapeutic potentials this miRNA upholds by silencing its oncogenic targets *CDK4* and *CDK6* and restoring its expression by epigenic demethylation to inhibit tumour cell growth and proliferation [82, 92–95].

There are several miRNAs expressed in medulloblastoma (**Table 1**) and they have been investigated over the last two decades aiming to identify novel biomarkers for diagnostic, prognostic and therapeutic purposes. The analysis of miRNAs as potential biomarkers is performed using MB tissue, CSF, and blood samples, in addition to the investigation of the miRNA expression in extracellular vesicles isolated from CSF or blood samples. Several miRNAs including miR-30b/d, miR-128a, miR-124, miR106b, and miR 224were found to be differentially expressed in MBs subgroups (**Table 1**) [82, 93].

7. Nanoparticles and miRNAs

In recent years, a novel approach for diagnosis, therapy and/or theranostics by the encapsulation of miRNAs in nanoparticles has emerged and it has been the centre of focus of several studies [15, 17, 97]. For instance, miR-124 and anti-miR-21 were co-encapsulated in polymeric nanocontainers that had a surface modified with Angiopep-2 peptide and injected in mice model. Results revealed promising outcomes in reducing the tumorigenesis of the glioblastoma in the xenograft mice model.

This approach also offered the protection of the encapsulated miRNA from enzyme degradation and assured the overcome of the BBB and the achievement of targeted dual delivery of miR-124 and anti-miR-21.

Co-encapsulation and delivery of the chemotherapeutic drug doxorubicin and anti-miR-21 showed significant decrease in miR-21 expression levels, reduction in tumour growth, and enhancement of the apoptotic activity *in vivo* [15]. Further examples that demonstrate the distinctive therapeutic potential of miRNAs loaded in nanocontainers are the gold nanoparticles functionalized with miR-182. Their administration intravenously in orthotopic glioblastoma xenografts resulted in significant antitumour activity and reduced tumour growth due to the protection of the miR182 and the targeted delivery approach by the nanoparticles. The studies' findings further showed no inflammatory responses related to the miR-182's systemic introduction and insignificant cytotoxicity levels or side effects [17, 96, 97]. Yet, no recent studies focusing on the delivery of miRNAs by the mean of nanoparticles in paediatric brain tumours and in medulloblastoma specifically have been reported [96].

8. Conclusion

The unsatisfying clinical outcomes associated with primary brain tumour malignancies have led to intensifying the work on understanding miRNAs. Their potential as biomarkers and as therapeutic targets could enable their incorporation within the early detection, prognosis and possible treatment of brain tumours. Despite the advances in the molecular techniques used to analyse the role of miRNAs in several cancers, and the plethora of miRNAs reported as potential biomarkers for cancer diagnosis or treatment, no miRNAs candidates have exceled to the Food and Drug Association approval process [92, 93]. Their clinical adaptation is still far from acceptance, due to uncertainties in the experimental findings. Thus, current research should concentrate on the clinical utilisation of miRNAs as potential novel diagnostic and therapeutic tools.

Conflict of interest

The authors declare no conflict of interest.

Author details

Denis Mustafov[1,2], Sara Seriah[1,3,4], Roozba Malik[1] and Maria Braoudaki[1]*

1 School of Life and Medical Sciences, University of Hertfordshire, UK

2 College of Health, Medicine, and Life Sciences, Brunel University London, UK

3 Department of Chemistry, Inorganic Chemistry Lab, National and Kapodistrian University of Athens, Greece

4 "NCSR" Demokritos, Sol-Gel Laboratory, Institute of Nanoscience and Nanotechnology, Greece

*Address all correspondence to: m.braoudaki@herts.ac.uk

References

[1] Lapointe S, Perry A, Butowski NA. Primary brain tumours in adults. Lancet. 2018;**392**(10145):432-446. DOI: 10.1016/S0140-6736(18)30990-5

[2] Cancer Research UK. Survival, Brain and Spinal Cord Tumours. 2022. Available from: https://www.cancerresearchuk.org/about-cancer/brain-tumours/survival. [Accessed: August 30, 2022]

[3] Cancer Research UK. Brain, Other CNS, and Intracranial Tumours Incidence Statistics. 2022. Available from: https://www.cancerresearchuk.org/health-professional/cancer-statistics/statistics-by-cancer-type/brain-other-cns-and-intracranial-tumours/incidence#heading-Four. [Accessed: August 30, 2022]

[4] National Brain Tumor Society. Quick Brain Tumor Facts. 2022. Available from: https://braintumor.org/brain-tumor-information/brain-tumor-facts/#pediatric-brain-tumors. [Accessed: August 30, 2022]

[5] Meyers C, Smith J, Bezjak A, Mehta M, Liebmann J, Illidge T, et al. Neurocognitive function and progression in patients with brain metastases treated with whole-brain radiation and motexafin gadolinium: Results of a randomized phase III trial. Journal of Clinical Oncology. 2004;**22**(1):157-165. DOI: 10.1200/JCO.2004.05.128

[6] Cancer.Net. Brain Tumor: Statistics. 2022. Available from: https://www.cancer.net/cancer-types/brain-tumor/statistics. [Accessed: August 30, 2022]

[7] Balachandran A, Larcher L, Chen S, Veedu R. Therapeutically significant microRNAs in primary and metastatic brain malignancies. Cancers (Basel). 2020;**12**(9):1-29. DOI: 10.3390/CANCERS12092534

[8] Salles D, Laviola G, Malinverni A, Stávale J. Pilocytic astrocytoma: A review of general, clinical, and molecular characteristics. Journal of Child Neurology. 2020;**35**(12):852-858. DOI: 10.1177/0883073820937225

[9] Mathupala S, Mittal S, Guthikonda M, Sloan A. MicroRNA and brain tumors: A cause and a cure? DNA and Cell Biology. 2007;**26**(5):301. DOI: 10.1089/DNA.2006.0560

[10] Pollack I, Agnihotri S, Broniscer A. Childhood brain tumours: Current management, biological insights, and future directions. Journal of Neurosurgery. Pediatrics. 2019;**23**(3):261-273. DOI: 10.3171/2018

[11] Millard N, De Braganca K. Medulloblastoma. Journal of Child Neurology. 2016;**31**(12):1341-1353. DOI: 10.1177/0883073815600866

[12] Ward E, Hopkins M, Arbuckle L, Williams N, Forsythe L, Bujkiewicz S, et al. Nutritional problems in children treated for medulloblastoma: Implications for enteral nutrition support. Pediatric Blood & Cancer. 2009;**53**(4):570-575. DOI: 10.1002/PBC.22092

[13] Mahapatra S, Amsbaugh M. Medulloblastoma. StatPearls; 2022. Available from: https://www.ncbi.nlm.nih.gov/books/NBK431069/. [Accessed: August 30, 2022]

[14] Chen M, Medarova Z, Moore A. Role of microRNAs in glioblastoma. Oncotarget. 2021;**12**(17):1707-1723. DOI: 10.18632/oncotarget.28039

[15] Haumann R, Videira J, Kaspers G, van Vuurden D, Hulleman E. Overview of current drug delivery methods across the blood–brain barrier for the treatment of primary brain tumours. CNS Drugs. 2020;**34**(11):1121-1131. DOI: 10.1007/S40263-020-00766-W

[16] O'Brien J, Hayder H, Zayed Y, Peng C. Overview of microRNA biogenesis, mechanisms of actions, and circulation. Frontiers in Endocrinology (Lausanne). 2018;**9**:402. DOI: 10.3389/FENDO.2018.00402/BIBTEX

[17] Kim Y, Kim V. Processing of intronic microRNAs. The EMBO Journal. 2007;**26**(3):775-783. DOI: 10.1038/SJ.EMBOJ.7601512

[18] Hanna J, Hossain G, Kocerha J. The potential for microRNA therapeutics and clinical research. Frontiers in Genetics. 2019;**10**:478. DOI: 10.3389/FGENE.2019.00478/XML/NLM

[19] Alarcón C, Lee H, Goodarzi H, Halberg N, Tavazoie S. N6-methyladenosine marks primary microRNAs for processing. Nature. 2015;**519**(7544):482-485. DOI: 10.1038/nature14281

[20] Denli A, Tops B, Plasterk R, Ketting R, Hannon G. Processing of primary microRNAs by the microprocessor complex. Nature. 2004;**432**(7014):231-235. DOI: 10.1038/nature03049

[21] Han J, Lee Y, Yeom K, Kim Y, Jin H, Kim V. The Drosha-DGCR8 complex in primary microRNA processing. Genes & Development. 2004;**18**(24):3016-3027. DOI: 10.1101/GAD.1262504

[22] Stavast C, Erkeland S. The non-canonical aspects of microRNAs: Many roads to gene regulation. Cell. 2019;**8**(11): 1-20. DOI: 10.3390/CELLS8111465

[23] Zhang H, Kolb F, Jaskiewicz L, Westhof E, Filipowicz W. Single processing Centre models for human dicer and bacterial RNase III. Cell. 2004;**118**(1):57-68. DOI: 10.1016/j.cell.2004.06.017

[24] Yoda M, Kawamata T, Paroo Z, Ye X, Iwasaki S, Liu Q, et al. ATP-dependent human RISC assembly pathways. Nature Structural & Molecular Biology. 2009;**17**(1):17-23. DOI: 10.1038/nsmb.1733

[25] Hawley ZCE, Campos-Melo D, Droppelmann CA, Strong MJ. MotomiRS: MiRNAs in motor neuron function and disease. Frontiers in Molecular Neuroscience. 2017;**10**:127. DOI: 10.3389/FNMOL.2017.00127/BIBTEX

[26] Babiarz J, Ruby J, Wang Y, Bartel D, Blelloch R. Mouse ES cells express endogenous shRNAs, siRNAs, and other microprocessor-independent, dicer-dependent small RNAs. Genes & Development. 2008;**22**(20):2773-2785. DOI: 10.1101/GAD.1705308

[27] Yang S, Maurin T, Robine N, Rasmussen K, Jeffrey K, Chandwani R, et al. Conserved vertebrate mir-451 provides a platform for dicer-independent, Ago2-mediated microRNA biogenesis. Proceedings of the National Academy of Sciences of the United States of America. 2010;**107**(34):15163-15168. DOI: 10.1073/PNAS.1006432107/SUPPL_FILE/SD03.XLS

[28] Cheloufi S, Dos Santos C, Chong M, Hannon G. A dicer-independent miRNA biogenesis pathway that requires ago catalysis. Nature. 2010;**465**(7298):584-589. DOI: 10.1038/nature09092

[29] Hayder H, O'Brien J, Nadeem U, Peng C. MicroRNAs: Crucial regulators of placental development. Reproduction.

2018;**155**(6):R259-R271. DOI: 10.1530/REP-17-0603

[30] O'Brien J, Nguyen V, Peng C. Overview on miRNA classification, biogenesis, and functions. MicroRNA in Human Malignancies. 2018;**9**:1-12. DOI: 10.1016/B978-0-12-822287-4.00027-X

[31] Anthiya S, Griveau A, Loussouarn C, Baril P, Garnett M, Issartel J, et al. MicroRNA-based drugs for brain tumours. Trends in Cancer. 2018;**4**(3):222-238. DOI: 10.1016/J.TRECAN.2017.12.008

[32] Petrescu G, Sabo A, Torsin L, Calin G, Dragomir M. MicroRNA based theranostics for brain cancer: Basic principles. Journal of Experimental & Clinical Cancer Research. 2019;**38**(1):1-21. DOI: 10.1186/S13046-019-1180-5

[33] Lu J, Getz G, Miska E, Alvarez-Saavedra E, Lamb J, Peck D, et al. MicroRNA expression profiles classify human cancers. Nature. 2005;**435**(7043):834-838. DOI: 10.1038/NATURE03702

[34] Xue P, Huang S, Han X, Zhang C, Yang L, Xiao W, et al. Exosomal miR-101-3p and miR-423-5p inhibit medulloblastoma tumorigenesis through targeting FOXP4 and EZH2. Cell Death and Differentiation. 2022;**29**(1):82-95. DOI: 10.1038/S41418-021-00838-4

[35] Yeh M, Wang Y, Yoo J, Oh C, Otani Y, Kang J, et al. MicroRNA-138 suppresses glioblastoma proliferation through downregulation of CD44. Scientific Reports. 2021;**11**(1):1-11. DOI: 10.1038/s41598-021-88615-8

[36] Costa P, Cardoso A, Custódia C, Cunha P, Pereira De Almeida L, Pedroso De Lima M. MiRNA-21 silencing mediated by tumor-targeted nanoparticles combined with sunitinib: A new multimodal gene therapy approach for glioblastoma. Journal of Controlled Release. 2015;**207**:31-39. DOI: 10.1016/J.JCONREL.2015.04.002

[37] Møller H, Rasmussen A, Andersen H, Johnsen K, Henriksen M, Duroux M. A systematic review of MicroRNA in glioblastoma multiforme: micro-modulators in the mesenchymal mode of migration and invasion. Molecular Neurobiology. 2012;**47**(1):131-144. DOI: 10.1007/s12035-012-8349-7

[38] Goenka A, Tiek D, Song X, Iglesia RP, Lu M, Hu B, et al. The role of non-coding RNAs in glioma. Biomedicine. 2022;**10**(8):2031. DOI: 10.3390/biomedicines10082031

[39] Slack F, Chinnaiyan A. The role of non-coding RNAs in oncology. Cell. 2019;**179**(5):1033-1055. DOI: 10.1016/j.cell.2019.10.017

[40] da Costa MP, De Windt L. miR-21: a miraculous Socratic paradox. Cardiovascular Research. 2010;**87**(3):397-400. DOI: 10.1093/cvr/cvq196

[41] Qu S, Guan J, Liu Y. Identification of microRNAs as novel biomarkers for glioma detection: A meta-analysis based on 11 articles. Journal of the Neurological Sciences. 2015;**348**(1-2):181-187. DOI: 10.1016/j.jns.2014.11.036

[42] Yang C, Yue J, Pfeffer S, Fan M, Paulus E, Hosni-Ahmed A, et al. microRNA-21 promotes glioblastoma tumorigenesis by down-regulating insulin-like growth factor-binding protein-3 (IGFBP3). Journal of Biological Chemistry. 2014;**289**(36):25079-25087. DOI: 10.1074/jbc.m114.593863

[43] ParvizHamidi M, Haddad G, Ostadrahimi S, Ostadrahimi N, Sadeghi S, Fayaz S, et al. Circulating miR-26a and

miR-21 as biomarkers for glioblastoma multiform. Biotechnology and Applied Biochemistry. 2019;**66**(2):261-265. DOI: 10.1002/bab.1707

[44] Jesionek-Kupnicka D, Braun M, Trąbska-Kluch B, Czech J, Szybka M, Szymańska B, et al. MiR-21, miR-34a, miR-125b, miR-181d and miR-648 levels inversely correlate with MGMT and TP53 expression in primary glioblastoma patients. Archives of Medical Science. 2019;**15**(2):504-512. DOI: 10.5114/aoms.2017.69374

[45] Ma C, Nguyen H, Luwor R, Stylli S, Gogos A, Paradiso Kaye A, et al. A comprehensive meta-analysis of circulation miRNAs in glioma as potential diagnostic biomarker. PLoS One. 2018;**13**(2):e0189452. DOI: 10.1371/journal.pone.0189452

[46] Ivo D'Urso P, Fernando D'Urso O, Damiano Gianfreda C, Mezzolla V, Storelli C, Marsigliante S. miR-15b and miR-21 as circulating biomarkers for diagnosis of glioma. Current Genomics. 2015;**16**(5):304-311. DOI: 10.2174/1389202916666150707155610

[47] Qu K, Lin T, Pang Q, Liu T, Wang Z, Tai M, et al. Extracellular miRNA-21 as a novel biomarker in glioma: Evidence from meta-analysis, clinical validation, and experimental investigations. Oncotarget. 2016;7(23):33994-34010. DOI: 10.18632/oncotarget.9188

[48] Ilhan-Mutlu A, Wagner L, Wöhrer A, Furtner J, Widhalm G, Marosi C, et al. Plasma microRNA-21 concentration may be a useful biomarker in glioblastoma patients. Cancer Investigation. 2012;**30**(8):615-621. DOI: 10.3109/07357907.2012.708071

[49] Mao X, Sun Y, Tang J. Serum miR-21 is a diagnostic and prognostic marker of primary central nervous system lymphoma. Neurological Sciences. 2013;**35**(2):233-238. DOI: 10.1007/s10072-013-1491-9

[50] Chuang H, Su Y, Liu H, Chen C, Chiu S, Cho D, et al. Preclinical evidence of STAT3 inhibitor pacritinib overcoming temozolomide resistance via downregulating miR-21-enriched exosomes from M2 glioblastoma-associated macrophages. Journal of Clinical Medicine. 2019;**8**(7):959. DOI: 10.3390/jcm8070959

[51] Liang T, Han L, Guo L. Rewired functional regulatory networks among miRNA isoforms (isomiRs) from let-7 and miR-10 gene families in cancer. Computational and Structural Biotechnology Journal. 2020;**13**(18):1238-1248. DOI: 10.1016/j.csbj.2020.05.001

[52] Teplyuk N, Uhlmann E, Wong A, Karmali P, Basu M, Gabriely G, et al. MicroRNA-10b inhibition reduces E2F1-mediated transcription and miR-15/16 activity in glioblastoma. Oncotarget. 2015;**6**(6):3770-3783. DOI: 10.18632/oncotarget.3009

[53] Siegal T, Charbit H, Paldor I, Zelikovitch B, Canello T, Benis A, et al. Dynamics of circulating hypoxia-mediated miRNAs and tumor response in patients with high-grade glioma treated with bevacizumab. Journal of Neurosurgery. 2016;**125**(4):1008-1015. DOI: 10.3171/2015.8.JNS15437

[54] Shi J. Considering exosomal miR-21 as a biomarker for cancer. Journal of Clinical Medicine. 2016;**5**(4):42. DOI: 10.3390/jcm5040042

[55] Olioso D, Caccese M, Santangelo A, Lippi G, Zagonel V, Cabrini G, et al. Serum exosomal microRNA-21, 222 and 124-3p as noninvasive predictive biomarkers in newly diagnosed high-grade gliomas: A prospective study.

Cancers (Basel). 2021;**13**(12):3006. DOI: 10.3390/cancers13123006

[56] Galardi S, Mercatelli N, Giorda E, Massalini S, Frajese G, Ciafre S, et al. miR-221 and miR-222 expression affects the proliferation potential of human prostate carcinoma cell lines by targeting p27Kip1. Journal of Biological Chemistry. 2007;**282**(32):23716-23724. DOI: 10.1074/jbc.M701805200

[57] Zhang R, Pang B, Xin T, Guo H, Xing Y, Xu S, et al. Plasma miR-221/222 family as novel descriptive and prognostic biomarkers for glioma. Molecular Neurobiology. 2015;**53**(3):1452-1460. DOI: 10.1007/s12035-014-9079-9

[58] Swellam M, Ezz El Arab L, Al-Posttany AS, Said B, S. Clinical impact of circulating oncogenic MiRNA-221 and MiRNA-222 in glioblastoma multiform. Journal of Neuro-Oncology. 2019;**144**(3):545-551. DOI: 10.1007/s11060-019-03256-2

[59] Li W, Guo F, Wang P, Hong S, Zhang C. miR-221/222 confers radioresistance in glioblastoma cells through activating Akt independent of PTEN status. Current Molecular Medicine. 2014;**14**(1):185-195. DOI: 10.2174/1566524013666131203103147

[60] Tokudome T, Sasaki A, Tsuji M, Udaka Y, Oyamada H, Tsuchiya H, et al. Reduced PTEN expression and overexpression of miR-17-5p, –19a-3p, –19b-3p, –21-5p, –130b-3p, –221-3p and –222-3p by glioblastoma stem-like cells following irradiation. Oncology Letters. 2015;**10**(4):2269-2272. DOI: 10.3892/ol.2015.3594

[61] Santangelo A, Imbrucè P, Gardenghi B, Belli L, Agushi R, Tamanini A, et al. A microRNA signature from serum exosomes of patients with glioma as complementary diagnostic biomarker. Journal of Neuro-Oncology. 2018;**136**(1):51-62. DOI: 10.1007/s11060-017-2639-x

[62] Bruno I, Karam R, Huang L, Bhardwaj A, Lou C, Shum E, et al. Identification of a microRNA that activates gene expression by repressing nonsense-mediated RNA decay. Molecular Cell. 2011;**42**(4):500-510. DOI: 10.1016/j.molcel.2011.04.018

[63] Smirnova L, Gräfe A, Seiler A, Schumacher S, Nitsch R, Wulczyn F. Regulation of miRNA expression during neural cell specification. European Journal of Neuroscience. 2005;**21**(6):1469-1477. DOI: 10.1111/j.1460-9568.2005.03978.x

[64] Novakova J, Slaby O, Vyzula R, Michalek J. microRNA involvement in glioblastoma pathogenesis. Biochemical and Biophysical Research Communications. 2009;**386**(1):1-5. DOI: 10.1016/j.bbrc.2009.06.034

[65] Roth P, Wischhusen J, Happold C, Chandran P, Hofer S, Eisele G, et al. A specific miRNA signature in the peripheral blood of glioblastoma patients. Journal of Neurochemistry. 2011;**118**(3):449-457. DOI: 10.1111/j.1471-4159.2011.07307.x

[66] Banelli B, Forlani A, Allemanni G, Morabito A, Pistillo M, Romani M. microRNA in glioblastoma: An overview. International Journal of Genomics. 2017;**2017**:1-16. DOI: 10.1155/2017/7639084

[67] Amini S, Abak A, Sakhinia E, Abhari A. microRNA-221 and microRNA-222 in common human cancers: Expression, eunction, and triggering of tumour progression as a key modulator. Laboratory Medicine. 2019;**50**(4):333-347. DOI: 10.1093/labmed/lmz002

[68] Nie Q, Lin Y, Yang X, Shen L, Guo L, Que SL, et al. IDH1R^{132}H decreases the proliferation of U87 glioma cells through upregulation of microRNA-128a. Molecular Medicine Reports. 2015;**12**(5):6695-6701. DOI: 10.3892/mmr.2015.4241

[69] Hu G, Fang W, Liu N, Li C. Effects of mir-128a on the invasion and proliferation of glioma U251 cells. Oncology Letters. 2019;**17**(1):891-896. DOI: 10.3892/ol.2018.9651

[70] Wang M, Li Q, Yu S, Zhang Z, Qiu P, Zhang Y, et al. Coronin 3 promotes the development of oncogenic properties in glioma through the Wnt/β-catenin signaling pathway. Oncotargets and Therapy. 2020;**7**(13):6661-6673. DOI: 10.2147/OTT.S257001

[71] Li Y, Guessous F, Zhang Y, DiPierro C, Kefas B, Johnson E, et al. microRNA-34a inhibits glioblastoma growth by targeting multiple oncogenes. Cancer Research. 2009;**69**(19):7569-7576. DOI: 10.1158/0008-5472.CAN-09-0529

[72] Gao H, Zhao H, Xiang W. Expression level of human miR-34a correlates with glioma grade and prognosis. Journal of Neuro-Oncology. 2013;**113**(2):221-228. DOI: 10.1007/s11060-013-1119-1

[73] Kipper F, Kieran M, Thomas A, Panigrahy D. Notch signaling in malignant gliomas: Supporting tumor growth and the vascular environment. Cancer Metastasis Reviews. 2022;**41**:27. DOI: 10.1007/s10555-022-10041-7

[74] Lam H-K, Ng P, di Tomaso E, Ng H, Pang J, Roussel M, et al. Expression of p19 INK4d, CDK4, CDK6 in glioblastoma multiforme. British Journal of Neurosurgery. 2000;**14**(1):28-32. DOI: 10.1080/02688690042870

[75] Vojdani S, Ghaderian S, Zali A, Rakhshan A, Oraee Yazdani S, Poursheikhani A, et al. Altered expression of EGFR and miR-34a derived from serum and tumoral tissue was associated with glioblastoma multiform. Experimental and Molecular Pathology. 2021;**121**:104655. DOI: 10.1016/j.yexmp.2021.104655

[76] Zhao J, Tao Y, Zhou Y, Qin N, Chen C, Tian D, et al. MicroRNA-7: A promising new target in cancer therapy. Cancer Cell International. 2015;**15**(1). DOI: 10.1186/s12935-015-0259-0

[77] Yin C, Kong W, Jiang J, Xu H, Zhao W. miR 7 5p inhibits cell migration and invasion in glioblastoma through targeting SATB1. Oncology Letters. 2019;**17**(2):1819-1825. DOI: 10.3892/ol.2018.9777

[78] Yang B, Ma YB, Chu SH. Silencing SATB1 overcomes temozolomide resistance by downregulating MGMT expression and upregulating SLC22A18 expression in human glioblastoma cells. Cancer Gene Therapy. 2018;**25**(11-12):309-316. DOI: 10.1038/s41417-018-0040-3

[79] Wang W, Dai L, Zhang S, Yang Y, Yan N, Fan P. Regulation of epidermal growth factor receptor signalling by plasmid-based microRNA-7 inhibits human malignant gliomas growth and metastasis in vivo. Neoplasma. 2013;**60**(03):274-283. DOI: 0.4149/neo_2013_036

[80] Kefas B, Godlewski J, Comeau L, Li Y, Abounader R, Hawkinson LJ, et al. microRNA-7 inhibits the epidermal growth factor receptor and the Akt pathway and is down-regulated in glioblastoma. Cancer Research. 2008;**68**(10):3566-3572. DOI: 10.1158/0008-5472.CAN-07-6639

[81] Gatto L, Franceschi E, Di Nunno V, Tosoni A, Lodi R, Brandes A. Liquid

biopsy in glioblastoma management: From current research to future perspectives. The Oncologist. 2021;**26**(10):865-878. DOI: 10.1002/onco.13858

[82] Bevacqua E, Farshchi J, Niklison-Chirou M, Tucci P. Role of microRNAs in the development and progression of the four medulloblastoma subgroups. Cancers. 2021;**13**(24):6323. DOI: 10.3390/cancers13246323

[83] Kozuka T, Omori Y, Watanabe S, Tarusawa E, Yamamoto H, Chaya T, et al. miR-124 dosage regulates prefrontal cortex function by dopaminergic modulation. Scientific Reports. 2019;**9**:1-14. DOI: 10.1038/s41598-019-38910-2

[84] Ghafouri-Fard S, Shoorei H, Bahroudi Z, Abak A, Majidpoor J, Taheri M. An update on the role of miR-124 in the pathogenesis of human disorders. Biomedicine & Pharmacotherapy. 2021;**135**:111198. DOI: 10.1016/j.biopha.2020.111198

[85] Han D, Dong X, Zheng D, Nao J. MiR-124 and the underlying therapeutic promise of neurodegenerative disorders. Frontiers in Pharmacology. 2020;**10**:1-9. DOI: 10.1523/jneurosci.0319-16.2016

[86] Higuchi F, Uchida S, Yamagata H, Abe-Higuchi N, Hobara T, Hara K, et al. Hippocampal MicroRNA-124 enhances chronic stress resilience in mice. Journal of Neuroscience. 2016;**36**(27):7253-7267. DOI: 10.1523/jneurosci.0319-16.2016

[87] Roy B, Lee E, Li T, Rampersaud M. Role of miRNAs in neurodegeneration: From disease cause to tools of biomarker discovery and therapeutics. Genes. 2022;**13**(3):425. DOI: 10.3390/genes13030425

[88] Zhao Y, Li K, Han X, Pan Y. The mechanism of non-coding RNAs in medulloblastoma (review). Oncology Letters. 2021;**22**(5):1-7. DOI: 10.3892/ol.2021.13019

[89] Wang M, Zhao Y, Yu Z, Zhang R, Li S, Zhang P, et al. Glioma exosomal microRNA-148a-3p promotes tumor angiogenesis through activating the EGFR/MAPK signaling pathway via inhibiting ERRFI1. Cancer Cell International. 2020;**20**:1-16. DOI: 10.1186/s12935-020-01566-4

[90] Li M, Deng Y, Zhang W. Molecular determinants of medulloblastoma metastasis and leptomeningeal dissemination. Molecular Cancer Research. 2021;**19**(5):743-752. DOI: org/10.1158/1541-7786.MCR-20-1026

[91] Besharat Z, Sabato C, Po A, Gianno F, Abballe L, Napolitano M, et al. Low expression of miR-466f-3p sustains epithelial to mesenchymal transition in sonic hedgehog medulloblastoma stem cells through Vegfa-Nrp2 signaling pathway. Frontiers in Pharmacology. 2018;**9**:9. DOI: 10.3389/fphar.2018.01281

[92] Singh M, Gupta R, Comez L, Paciaroni A, Rani R, Kumar V. BCL2 G quadruplex-binding small molecules: Current status and prospects for the development of next-generation anticancer therapeutics. Drug Discovery Today. 2022;**27**(9):2551-2561. DOI: 10.1016/j.drudis.2022.06.002

[93] Joshi P, Katsushima K, Zhou R, Meoded A, Stapleton S, Jallo G, et al. The therapeutic and diagnostic potential of regulatory noncoding RNAs in medulloblastoma. Neuro-Oncology Advances. 2019;**1**(1):1-14. DOI: 10.1093/noajnl/vdz023

[94] Ivanov D, Coyle B, Walker D, Grabowska A. In vitro models of

medulloblastoma: Choosing the right tool for the job. Journal of Biotechnology. 2016;**236**:10-25. DOI: 10.1016/j.jbiotec.2016.07.028

[95] Cho Y, Tsherniak A, Tamayo P, Santagata S, Ligon A, Greulich H, et al. Integrative genomic analysis of Medulloblastoma identifies a molecular subgroup that drives poor clinical outcome. Journal of Clinical Oncology. 2011;**29**(11):1424-1430. DOI: 10.1200/jco.2010.28.5148

[96] Tantawy M, Elzayat M, Yehia D, Taha H. Identification of microRNA signature in different pediatric brain tumors. Genetics and Molecular Biology. 2018;**41**(1):27-34. DOI: 10.1590/1678-4685-gmb-2016-0334

[97] Venkataraman S, Alimova I, Fan R, Harris P, Foreman N, Vibhakar R. MicroRNA 128a increases intracellular ROS level by targeting Bmi-1 and inhibits Medulloblastoma Cancer cell growth by promoting senescence. PLoS One. 2010;**5**(6):e10748. DOI: 10.1371/journal.pone.0010748

[98] Yang S, Choi S, Lee J, Park A, Wang K, Phi J, et al. miR-192 suppresses leptomeningeal dissemination of medulloblastoma by modulating cell proliferation and anchoring through the regulation ofDHFR, integrins, andCD47. Oncotarget. 2015;**6**(41):43712-43730. DOI: 10.18632/oncotarget.6227

[99] Kanchan R, Perumal N, Atri P, Chirravuri Venkata R, Thapa I, Klinkebiel D, et al. MiR-1253 exerts tumor-suppressive effects in medulloblastoma via inhibition of CDK6 and CD276 (B7-H3). Brain Pathology. 2020;**30**(4):732-745. DOI: 10.1111/bpa.12829

[100] Liu Y, Zheng M, Jiao M, Yan C, Xu S, Du Q, et al. Polymeric nanoparticle mediated inhibition of miR-21 with enhanced miR-124 expression for combinatorial glioblastoma therapy. Biomaterials. 2021;**276**:121036. DOI: 10.1016/j.biomaterials.2021.121036

Chapter 3

DNA Methylation in Cancer Epigenetics

Ewelina A. Klupczyńska

Abstract

DNA methylation is one of the most important epigenetic modifications next to acetylation or histone modifications, as it has a role in the homeostatic control of the cell and is strongly involved in the control of genome expression. DNA methylation, which is catalyzed by DNA methyltransferases (DNMTs), is one of the primary epigenetic mechanisms that control cell proliferation, apoptosis, differentiation, cell cycle, and transformation in eukaryotes. Hypomethylation and hypermethylation result in the activation or repression of genes and in a normal cell there is a strict balance between these processes. Abnormal DNA methylation is a well-known feature of cancer development and progression and can turn normal stem cells into cancer stem cells. Studies clearly show that DNA methylation regulates gene transcription functions in cancer pathogenesis. In cancer cells, DNA methylation patterns are largely modified, and therefore, methylation is used to distinguish cancer cells from normal, healthy cells. However, the mechanisms underlying changes in DNA methylation remain unexplored. However, it is known that oxidative stress (OS) is a key mechanism of carcinogenesis, and DNA methylation of genes that are active at OS may play a role in cancer development. Studies also show that DNA methylation is mediated by long noncoding RNA (lncRNA) under both physiological and pathological conditions. How cell-specific DNA methylation patterns are established or disrupted is a key question in developmental biology and cancer epigenetics.

Keywords: DNA methylation, epigenetics, cancer, oxidative stress, lncRNA, biomarkers

1. Introduction

Epigenetics in the etiology of cancer progression is of major importance. Epigenetic changes, such as histone modifications, DNA methylation, chromatin remodeling, nucleosome positioning, regulation by noncoding RNAs and precisely microRNAs, play a significant role in the cancerogenesis of different cancer types. Epigenetic processes result in altered levels of gene transcriptional activity without directly affecting the primary DNA nucleotide sequence. Changes in DNA methylation patterns together with specific histone modifications (methylations, acetylations, deacetylations, etc.) contribute to a transcriptionally inactive chromatin state. In cancer cells, DNA methylation patterns are modified, and these differences are used in the diagnostic process to distinguish cancer cells from normal tissues [1].

 IntechOpen

Aberrations of normal DNA methylation patterns are observed in many cancers and are associated with chromatin alterations, changes in gene expression, and genome instability, making the study of DNA methylation crucial to understanding cancer biology and evolution and biomarker development [2]. Epigenetic clocks, assessed by DNA methylation levels, are among the most commonly used biological age markers in cancer research [3], as DNA methylation occurs early in tumorigenesis and often precedes somatic cell mutation [4].

2. DNA methylation in cancer

DNA methylation, as an epigenetic mechanism, occurs through the addition of a methyl group at the 5′ position of the pyrimidine ring of cytosines and plays an important role in cellular function, particularly in the transcriptional regulation of embryonic and adult stem cells. Genomewide analysis revealed different patterns of DNA methylation in different cell types, developmental stages, and in response to different stimuli. Abnormal DNA methylation patterns—hypomethylation or hypermethylation—cause gene expression or inhibition, lead to genome instability and DNA breakage. DNA hypomethylation occurs in the intergenic regions and repetitive DNA sequences of cancer cells, while DNA hypermethylation occurs at CpG islands in gene promoters. DNA hypermethylation is mediated by DNA methyltransferases, DNMT3a, 3b, and DNMT1, during *de novo* methylation (**Figure 1**). There is a tight balance between the regulation of gene activation or repression in normal cellular activity. When this balance is disrupted, for example, by oxidative stress, anomalous states arise. Hypomethylation promotes genomic instability, causing erroneous aggregation of chromosomes during cell division and unwanted activation of transposable elements in the genome, leading to further genetic damage [5, 6]. DNA hypomethylation can lead to activation of oncogenes (e.g. mesothelin, proopiomelanocortin gene, S100A4, and claudin4) and cancer progression, while DNA hypermethylation can lead to inactivation of tumor-suppressor

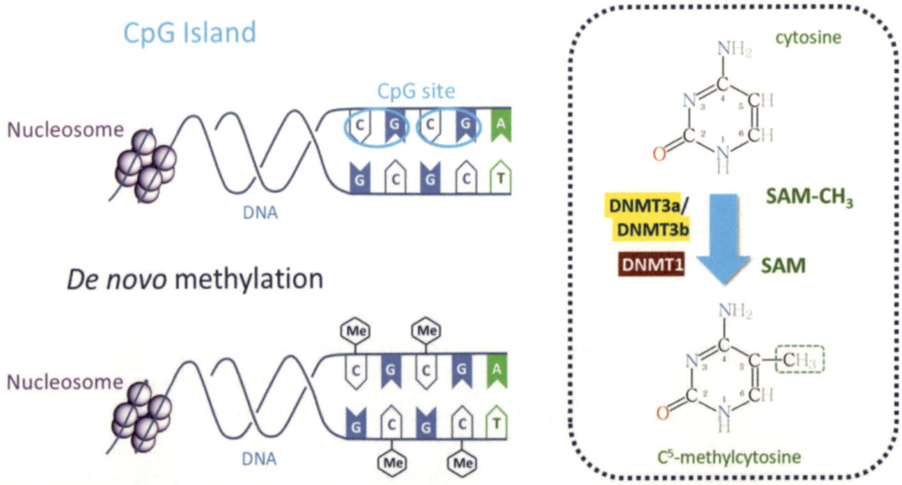

Figure 1.
DNA methylation and DNA methyltransferases (DNMT) function. DNA hypomethylation occurs in the intergenic regions of cancer cells, while DNA hypermethylation at CpG islands occurs in gene promoters. DNA methyltransferases: DNMT 1, DNMT3a, and DNMT3b play a major role in the methylation process.

genes, resulting in deregulated cell growth or altered response to anticancer therapies [7, 8]. DNA hypermethylation inactivates tumor-suppressor genes such as adenomatous polyposis coli (APC), retinoblastoma (Rb), or BRCA1, as well as others involved in DNA repair such as MGMT, apoptosis (DAPK) or antioxidation (GSTP1) [8, 9]. Abnormal methylation turns normal stem cells into cancer stem cells (CSCs). CSCs are small populations of cancer cells that exhibit unique properties such as self-regeneration, resistance to chemotherapy, and high metastatic capacity [10].

2.1 Effect of DNA methylation on cancer progression

Epigenetic changes, particularly changes in DNA methylation, are much more common than the frequency of genetic changes in the vast majority of cancer types. This is significant and demonstrates the importance of epigenetics both when used as a diagnostic and therapeutic tool.

The expression of DNA methyltransferases (DNMTs), which are responsible for carrying the methyl group during methylation, gradually increases with the transformation process from normal tissue to precancerous lesions and becomes overexpressed in cancer cells. In humans, during early embryogenesis, DNMT1, DNMT3a, and DNMT3b are the enzymes that establish methylation patterns. Abnormal DNA methylation—hypermethylation and hypomethylation—is closely associated with cancer tumor development. DNMT overexpression leads to hypermethylation of gene promoters, DNA methylation of the tissue-specific gene, thus hypermethylation of the CpG island can lead to tumor development (**Figure 2**). The most common genes with hypomethylation are oncogenes whose expression is upregulated in tumor progression. Therefore, increasing evidence has revealed specific mechanisms of methyltransferases (writer), demethylases (eraser), and DNA-binding proteins (reader) in the regulation of abnormal methylation during tumor development [4]. Demethylases remove the methyl group as an eraser, while readers are a class of proteins that are able to recognize the methylation mark by their distinct domains and induce different biological functions.

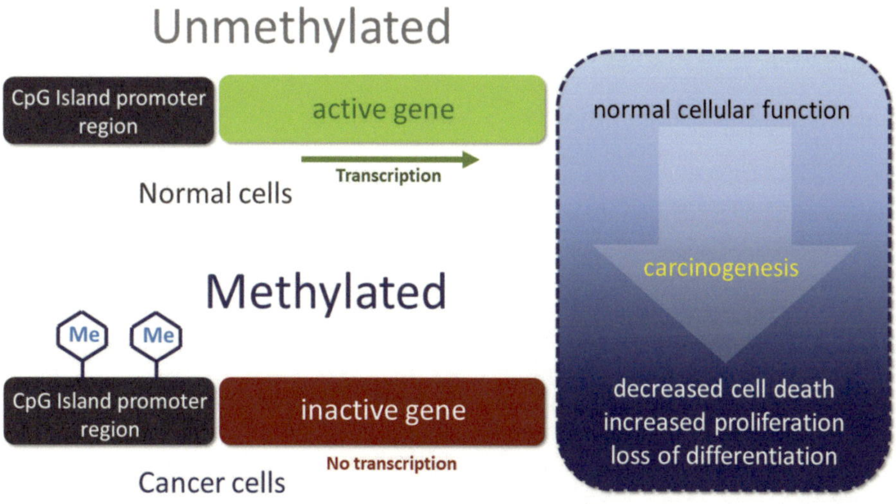

Figure 2.
Effect of DNA methylation on carcinogenesis.

Modification of DNA methylation affects a number of biological processes, such as growth, differentiation, and transformation mechanisms of eukaryotic cells. In mammals, it plays the most important role in placental and embryonic development and is essential for embryonic development. Abnormal DNA methylation can lead to many disorders in the body. DNA methylation is associated with diseases such as autoimmune diseases, neurodegenerative diseases such as Parkinson's and Alzheimer's disease, rheumatoid arthritis or various types of cancer. Studies have confirmed abnormal DNA methylation of fixed loci in many types of cancers, such as colon, pancreatic, breast, ovarian, esophageal, bladder, kidney, or bone cancer.

2.2 Oxidative stress in DNA methylation

Oxidative stress (OS) is the primary mechanism of cancerogenesis, and DNA methylation regulates gene transcription functions in cancer pathogenesis [11]. We speak of oxidative stress when there is excessive production of reactive oxygen species (ROS) in the cell and when the cell has a reduced antioxidant defense. A potentially mutagenic change in ROS-induced DNA damage is O6-methylguanine. Many studies have shown that its presence can inhibit the binding of DNA methyltransferases, that is, it can lead to hypomethylation by inhibiting the methylation of neighboring cytosine molecules. Alternatively, O6-methylguanine may spontaneously misuse thymine and thus contribute to DNA hypomethylation. Studies have shown that genomewide hypomethylation increases mutation rates and thus leads to genome instability. In addition, single-stranded DNA may signal *de novo* methylation, so it may be possible that the formation of single strand breaks by oxidative stress may contribute to the modification of DNA methylation patterns observed in oxidant-transformed cell lines.

Cancer is a multistep process and often involves changes in the transcriptional activity of genes associated with many critical cellular processes for tumor development, such as proliferation, aging, inflammation, or metastasis. Both genotoxic and nongenotoxic mechanisms contribute to malignant transformation. Genotoxic mechanisms include changes in genomic DNA sequences that ultimately lead to mutations. Nongenotoxic include mechanisms (other than those directly affecting DNA) capable of modulating gene expression. ROS have been implicated at all stages of the cancerogenic process through the involvement of both types of mechanisms.

2.3 LncRNA in DNA methylation

Histone modification and chromosome remodeling, as well as transcription factors, play a key role in regulating DNA methylation across the genome in a site-specific manner. The human genome contains thousands of noncoding regions that for decades were considered "junk DNA" due to the lack of evidence for their transcription and lack of protein coding. In humans, less than 2% of the genome encodes proteins, but studies have shown that the genome, including noncoding regions, can be actively transcribed into noncoding RNAs (up to 75%). RNA transcripts >200 nucleotides that do not encode proteins are described as lncRNAs. According to version 42 of GENCODE, 19933 *lncRNA* genes and 57,936 lncRNA loci transcripts have been identified in the human genome. LncRNAs can not only affect gene expression itself but also regulate gene expression and various signaling pathways by, among other things, interacting with DNA methylation. LncRNAs can act both in the nucleus and in the cytoplasm. In the nucleus, lncRNAs regulate chromatin remodeling and transcription, and in the cytoplasm, lncRNAs regulate mRNA translation and turnover. They can interfere with signaling pathways, many of which will affect gene expression in a variety of biological and pathophysiological conditions. LncRNAs have been committed in the acquisition of all features of tumor cells, from intrinsic proliferation and survival capacity, to increased metabolism and association with the tumor microenvironment.

DNA methylation is variously mediated by lncRNAs, regulating the expression of target genes in many processes, including pathological ones. LncRNAs can recruit or repel DNA methyltransferases and TETs, control SAM/SAH levels to regulate DNMT activity, and regulate the expression of DNMTs and TETs [12]. LncRNAs play a versatile role in development and in various disease processes, including carcinogenesis. Through next-generation sequencing, thousands of lncRNAs have been identified as abnormally altered in cancer tissues. By regulating DNA methylation, lncRNAs can have a major impact on malignant transformation and tumor progression—proliferation, invasion metastasis, and in addition, abnormal DNA methylation regulates the expression of lncRNAs as tumor-suppressor genes [13]. Studies also show that the lung cancer-associated lncRNA LUCAT1 has been implicated in the development of many cancers, such as clear cell renal cell carcinoma, nonsmall cell lung cancer, glioma, osteosarcoma, colorectal cancer, and gastric cancer [13]. The same is true for the lncRNAs HOTAIR, PCAT1, MALAT1, and FAL1, which are involved in various human cancers [14]. However, most *lncRNA* genes, especially cancer-related lncRNAs, need to be annotated and further studied.

The mechanism of action of lncRNAs and the role of lncRNAs in DNA methylation may offer prospects for the development of novel cancer drugs, and furthermore, as shown by lncRNAs themselves, can be used to develop new agents for the early diagnosis of, for example, gastric cancer with a sensitivity close to 80% [13]. The US Food and Drug Administration have approved the testing of patient urine samples for lncRNA PCA3 to detect prostate cancer. Further lncRNA biomarkers have been identified in hepatocellular carcinoma—lncRNA HULC [15] and gastric cancer—lncRNA HULC and CCAT, which is also associated with colon cancer [16].

3. DNA methylation in selected cancer types and biomarkers

Whole-genome hypomethylation is one of the first abnormal methylation events to alter the methylation signature of cells in many cancers [17]. Hypermethylation of the *APC* gene promoter, a tumor-suppressor gene, can cause abnormal cell proliferation, cell migration, cell adhesion, cytoskeletal reorganization, and chromosome stability in various cancer types [18].

3.1 Biomarkers in gastric cancer and skin cancer

Similarly, in gastric cancer (GC), DNA hypermethylation of tumor-suppressor genes and DNA hypomethylation of oncogenes tend to induce multistep carcinogenesis [13]. For GC, more than 100 such genes have been found, and about 70 genes are significantly hypermethylated in tumor tissues compared to genes observed in normal tissues in GC patients. In GC, one of the most important suppressor genes is E-cadhedrin, hMLH1 or APC [19]. E-cadhedrin contributes to tumor progression by increasing proliferation, invasion, and metastasis [19]. Studies show that promoter methylation of genes such as RUNX3, RASSF1A, and Reprimo is much more frequent in GC tissues compared to normal tissues, indicating that promoter methylation of these genes may induce cancer tumorigenesis [20]. Gastric cancer is a very common malignancy, often diagnosed at an advanced stage and with a poor prognosis. GC formation and progression are associated with epigenetic modifications such as DNA methylation, chromatin remodeling, posttranslational modifications of histones or noncoding RNAs. MGMT promoter methylation is also associated with GC risk, but may not be a potential biomarker for GC [20]. The situation is different for melanoma (SC). For SC, hypermethylation of the MGMT promoter is associated with a significantly increased risk of disease, as is methylation of the RAR-β2 promoter [21]. However, as shown in a meta-analysis on SC, the best biomarker for early detection of SC is RASSF1A, whose methylation was significantly associated with melanoma risk, and its loss is associated with SC pathogenesis, and reduced RASSF1A expression is correlated with hypermethylation of the CpG island promoter region [21].

3.2 Biomarkers in colorectal cancer

Also in colorectal cancer (CRC), promoter hypermethylation by affecting key cellular pathways such as DNA repair, apoptosis, cell cycle regulation or angiogenesis is associated with silencing of tumor-suppressor genes [22, 23]. Abnormal DNA methylation also occurs during the transformation of chronic inflammation into CRC. Long-term intestinal inflammation leads to errors in immune surveillance mechanisms, contributing to the fact that anti-tumor immune responses are inhibited, which in turn leads to tumor progression [24]. Furthermore, DNA methylation is an endogenous mutation generator, increasing DNA damage and thus contributing to cell apoptosis. This fact shows that DNA hypermethylation is also related to age. There are also hypermethylated tumor-suppressor genes among age-dependent genes, such as estrogen receptor-1 (ESR1), SFRP1, or SYNE1 [25]. There are also other well-studied methylated genes in colorectal cancer, such as vimentin (VIM), cadherin-1 (CDH1), MLH1, TIMP metallopeptidase inhibitor-3 (TIMP3), secreted frizzled-related protein-1 (SFRP1), and hypermethylated in cancer-1 (HIC1) [24]. Studies show that there are two biomarkers already introduced for the noninvasive diagnosis of CRC, methylated syndecan-2 (mSDC2), and methylated SEPT9 (mSEPT9) [26, 27].

3.3 Biomarkers in endometrial cancer

Abnormal DNA methylation in certain tumor-suppressor genes and oncogenes is also responsible for the process of carcinogenesis in the endometrium. Endometrial cancer (EC) is a malignant tumor of the female genital tract that is one of the most common worldwide. The incidence of EC is steadily increasing and is higher in developed countries. Epigenetic mechanisms also play an important role in the development and progression of EC, and changes in DNA methylation are one of the most important epigenomic modifications that play a role in EC development [28]. Studies on EC have shown that promoter methylation of the suppressor gene *RASSF1* is more frequent than in normal endometrium, and thus its expression is significantly reduced [29, 30]. Furthermore, RASSF1 methylation is also strongly correlated with EC risk and progression [31]. Multigene hypermethylation studies indicate that RASSF1 may be a potential biomarker in EC, as it is an important indicator of EC with high sensitivity and specificity [31]. It is rare in normal tissues, and RASSF1 promoter methylation is the most frequently inactivated gene identified among different cancer types.

3.4 Biomarkers in ovarian cancer, pancreatic cancer, and clear cell renal cell carcinoma

Studies of ovarian cancer (OC) and pancreatic cancer (PC) have shown that their progression is also linked to the accumulation of epigenetic changes. Women with advanced-stage OC have a five-year survival rate of less than 25%. Studies of DNA methylation markers have shown that, for OC cells, highly correlated with pathological fractions of OC cancer cells are, among others, the *ZNF154* gene [32] and for PC, the *SIM1*, *MIR129–2*, and *NR1I2* genes, which are specifically methylated in PC cells [33]. For OC, much attention has been paid to BRCA1 promoter methylation, as BRCA1 mutations are involved in hereditary OC. BRCA1 promoter hypermethylation occurs in 15–30% of cases of this cancer [34]. Nevertheless, studies indicate that ZNF154 methylation may serve better as a biomarker to detect OC and may also be a method capable of detecting multiple cancer types [35]. In the case of kidney cancer—clear cell renal cell carcinoma (ccRCC), such a biomarker appears to be hypermethylation of ZNF677 and PCDH8 [36].

DNA methylation profiling of tumor tissues is a valuable diagnostic tool for many types of cancer. DNA methylation data have become a valuable source of information for biomarker development. The discovery of functional and prognostic markers of DNA methylation in cancer provides both broader clinical opportunities and aids in the further development of epigenetic therapies. In addition, the DNA methylation data collected in The Cancer Genome Atlas (TCGA) helps to predict, for example, transcription factor (TF) regulators causing abnormal DNA methylation in different types of cancer, providing further therapeutic opportunities. Based on thousands of cancer samples of different types, TCGA collects, describes, and analyses data at clinical, molecular, and imaging levels, which are available in the Genomic Data Commons. The nature of DNA methylation aberrations in cancer and the stability of cell-free DNA in body fluids are of interest for the advancement of cancer diagnostics and therapy. The use of multiple target loci per test and genomewide epigenetic changes may help to improve the quality of tests under development (better sensitivity and specificity). Epigenetic modifications, related to tumorigenesis and cancer biology, as biomarkers specifically based on DNA methylation, due to the dynamic and reversible nature of this process, help to treat and improve therapies not only for cancer but also for other diseases (**Table 1**) [37].

4. Nutri-epigenomics in cancer

Epigenetic biomarkers, in particular, not only DNA methylation but also histone modifications or miRNA regulation are induced by environmental factors, including through dietary habits and dietary components. Studies show that diet modulates DNA methylation [38] and has a significant contribution to the tumor-suppressive potential of tumors, at initiation, promotion, and progression stages [39]. Drugs usually act on single targets, whereas dietary components have multidirectional effects on tumors [39]. Studies have shown that some bioactive dietary components act as methyl donors or methylation cofactors or modifiers of DNMT enzymatic activity [8]. Diet and bioactive compounds in food (i.e. vitamins) have chemopreventive effects,

Type of cancer	Biomarkers	References
Lung cancer (LC)	APC gene promoter methylation	[18]
Breast cancer (BC)	APC gene promoter methylation	[18]
Gastric cancer (GC)	RUNX3, Reprimo and RASSF1A promoter methylation	[20]
Skin cancer (SC), melanoma	DNA methylation of tumor-suppressor gene RASSF1A	[21]
Colorectal cancer (CRC)	Methylated syndecan-2 (mSDC2) and methylated SEPT9 (mSEPT9)	[26, 27]
Endometrial cancer (EC)	Promoter methylation of the suppressor gene RASSF1	[29, 30]
Ovarian cancer (OC)	ZNF154, BRCA1	[32, 34]
Pancreatic cancer (PC)	Methylated genes: *SIM1, MIR129–2 i NR1I2, ZNF154*	[33, 35]
Kidney (Renal Cell) Cancer	Hypermethylation of ZNF677 and PCDH8	[36]

Table 1.
Biomarkers for some of the most common cancers based on DNA methylation.

able to act as epigenetic modifiers in tumor cells, disrupting the molecular mechanisms responsible for inappropriate DNA methylation patterns. Disregulation of epigenetic patterns, on the other hand, disrupts gene expression and can be the cause of tumorigenesis and other diseases [40]. To prevent and treat CRC, for example, the use of vitamins in combination with DNA methyltransferase inhibitors and other approved therapies seems promising [40, 41]. Nevertheless, studies show that dietary change and active lifestyle alone are already associated with altered DNA methylation patterns, in DNA regions with gene functions related to immune cell metabolism, tumor suppression, and general aging [38].

Nutri-epigenomics is a promising field that is rapidly developing, based on the growing knowledge of DNA methylation and its interactions dependent on bioactive nutrients and their action as epigenetic modifiers with implications for anti-cancer therapies.

5. Perspectives

Cancer epigenetics is pioneering potential applications of epigenetics in clinical treatment with epigenetics-based biomarkers successfully demonstrated in cancer diagnosis, prediction of tumor progression, and prediction of therapeutic response. Epigenetic biomarkers also have the potential to be used as screening tools. The role of DNA methylation and its abnormalities in cancer has been well documented in scientific studies.

Through DNA methylation profiling in tumor tissues, new entities have been identified or morphologically distinct cancers have been combined into appropriate entities, classifying cancers accordingly. Most importantly, however, DNA methylation patterns help to a great extent in the detection of even minimally invasive tumor tissue and are increasingly being used as a good diagnostic tool for many types of cancer. Unlike genetic alterations, DNA methylation is reversible, which makes it of great interest for therapeutic approaches. However, more information and research is still needed on the subject, and we may be able to prevent or cure cancer in the future.

Although great progress has been made in understanding the role of DNA methylation, and hypermethylated promoters serve as biomarkers, and only a few methylated genes or functional elements serve as clinically relevant cancer biomarkers. The bottleneck in the progress of DNA methylation has shifted from data generation to data analysis. Therefore, the next step is to develop machine-learning models for computational estimation of methylation profiling and identification of potential biomarkers and to create algorithms to predict DNA methylation modifications [42, 43]. A thorough understanding of DNA methylation mechanisms and prediction of DNA methylation modifications will be more effective to use for cancer diagnosis and therapy.

6. Conclusions

DNA methylation is a specifically regulated biochemical process, an epigenetic modification that plays a key role in human development and is important for the homeostatic control of the cell. During DNA methylation, epigenetic aberrations replace normal cellular signals that lead to tumor initiation and propagation. DNA methylation profiling follows the continuous technological improvements of DNA methylation assays, which in turn can provide an enormous amount of data. It is also an important aspect of understanding malignant transformation and is becoming an increasingly important tool for diagnosing cancer (increasing the accuracy of diagnosis), predicting prognosis, and monitoring therapy. With an increasingly accurate and comprehensive understanding of the mechanism based on DNA methylation, better and better drugs can be created that target the epigenetic machinery of the cell as anti-cancer therapies. However, despite significant advances in the knowledge of DNA methylation in translational research, many challenges remain. Having understood the role of DNA methylation, the next challenge facing the community is to decipher the role that DNA methylation derivatives play in cancer. We have also only just begun to fully understand the genomewide demethylation process and the impact that methylation intermediates have on tumor development, diagnosis, and treatment. Expanding our knowledge of cancer treatment strategies will be greatly influenced by the growing field of epitranscriptomics, or understanding aberrations in DNA methylation of retroviruses. Thanks to projects such as TCGA or the Cancer Cell Line Encyclopedia, which provide genomic, epigenomic and transcriptomic data, early cancer detection therapy combined with personalized treatment is no longer just a dream.

Acronyms and abbreviations

APC	adenomatous polyposis coli
BC	breast cancer
CSC	cancer stem cells
CRC	colorectal cancer
CGI	CpG islands
DAPK	death-associated protein kinase-1
bp	DNA base pairs
DNA	deoxyribonucleic acid
DNMT	DNA methyltransferase

EC	endometrial cancer
GC	gastric cancer
GSTP1	glutathione S-Transferase P-1
lncRNA	long noncoding RNA
LC	lung cancer
OC	ovarian cancer
OS	oxidative stress
MGMT	O-6-Methylguanine-DNA Methyltransferase
PC	pancreatic cancer
ROS	reactive oxygen species
SAH	S-adenosyl-L-homocysteine
SAM	S-adenosyl-L-methionine
SC	skin cancer
TCGA	The Cancer Genome Atlas
TET	ten-eleven translocation
TF	transcription factor
TRDMT1	tRNA methyltransferase

Author details

Ewelina A. Klupczyńska
Polish Academy of Sciences in Kórnik, Poland

*Address all correspondence to: evelin@man.poznan.pl

References

[1] Delpu Y, Cordelier P, Cho WC, Torrisani J. DNA methylation and Cancer diagnosis. International Journal of Molecular Sciences. 2013;**14**(7):15029-15058

[2] Merkel A, Esteller M. Experimental and bioinformatic approaches to studying DNA methylation in cancer. Cancers. 2022;**14**(2):349

[3] Lau CHE, Robinson O. DNA methylation age as a biomarker for cancer. International Journal of Cancer. 2021;**148**(11):2652-2663

[4] Zhao Y, Yang M, Wang S, Abbas SJ, Zhang J, Li Y, et al. An overview of epigenetic methylation in pancreatic cancer progression. Frontiers in Oncology. 2022;**2022**:12. https://www.frontiersin.org/articles/10.3389/fonc.2022.854773

[5] Prada D, González R, Sánchez L, Castro C, Fabián E, Herrera LA. Satellite 2 demethylation induced by 5-azacytidine is associated with missegregation of chromosomes 1 and 16 in human somatic cells. Mutation Research/Fundamental and Molecular Mechanisms of Mutagenesis. 2012;**729**(1):100-105

[6] Daskalos A, Nikolaidis G, Xinarianos G, Savvari P, Cassidy A, Zakopoulou R, et al. Hypomethylation of retrotransposable elements correlates with genomic instability in non-small cell lung cancer. International Journal of Cancer. 2009;**124**(1):81-87

[7] Stone A, Zotenko E, Locke WJ, Korbie D, Millar EKA, Pidsley R, et al. DNA methylation of oestrogen-regulated enhancers defines endocrine sensitivity in breast cancer. Nature Communications. 2015;**6**(1):7758

[8] Mahmoud AM, Ali MM. Methyl donor micronutrients that modify DNA methylation and Cancer outcome. Nutrients. 2019;**11**(3):608

[9] Sproul D, Meehan RR. Genomic insights into cancer-associated aberrant CpG island hypermethylation. Briefings in Functional Genomics. 2013;**12**(3):174-190

[10] Mazloumi Z, Farahzadi R, Rafat A, Dizaji Asl K, Karimipour M, Montazer M, et al. Effect of aberrant DNA methylation on cancer stem cell properties. Experimental and Molecular Pathology. 2022;**125**:104757

[11] Gao T, Joyce BT, Liu L, Zheng Y, Dai Q, Zhang Z, et al. DNA methylation of oxidative stress genes and cancer risk in the normative aging study. American Journal of Cancer Research. 2016;**6**(2):553-561

[12] Huang W, Li H, Yu Q, Xiao W, Wang DO. LncRNA-mediated DNA methylation: An emerging mechanism in cancer and beyond. Journal of Experimental & Clinical Cancer Research. 2022;**41**(1):100

[13] Li N, Zeng A, Wang Q, Chen M, Zhu S, Song L. Regulatory function of DNA methylation mediated lncRNAs in gastric cancer. Cancer Cell International. 2022;**22**(1):227

[14] Hu X, Sood AK, Dang CV, Zhang L. The role of long noncoding RNAs in cancer: The dark matter matters. Current Opinion in Genetics & Development. 2018;**48**:8-15

[15] Panzitt K, Tschernatsch MMO, Guelly C, Moustafa T, Stradner M, Strohmaier HM. Characterization of

HULC, a novel gene with striking up-regulation in hepatocellular carcinoma, as noncoding RNA. Gastroenterology. 2007;**132**(1):330-342

[16] Shao Y, Ye M, Jiang X, Sun W, Ding X, Liu Z, et al. Gastric juice long noncoding RNA used as a tumor marker for screening gastric cancer. Cancer. 2014;**120**(21):3320-3328

[17] Detilleux D, Spill YG, Balaramane D, Weber M, Bardet AF. Pan-cancer predictions of transcription factors mediating aberrant DNA methylation. Epigenetics & Chromatination. 2022;**15**(1):10

[18] Michalak EM, Burr ML, Bannister AJ, Dawson MA. The roles of DNA, RNA and histone methylation in ageing and cancer. National Review in Molecular Cell Biology. 2019;**20**(10):573-589

[19] Nakamura J, Tanaka T, Kitajima Y, Noshiro H, Miyazaki K. Methylation-mediated gene silencing as biomarkers of gastric cancer: a review. World Journal of Gastroenterology. 2014;**20**(34):11991-12006

[20] Wen J, Zheng T, Hu K, Zhu C, Guo L, Ye G. Promoter methylation of tumor-related genes as a potential biomarker using blood samples for gastric cancer detection. Oncotarget. 2017;**8**(44):77783-77793

[21] Guo Y, Long J, Lei S. Promoter methylation as biomarkers for diagnosis of melanoma: A systematic review and meta-analysis. Journal of Cell Physiology. 2019;**234**(5):7356-7367

[22] Casalino L, Verde P. Multifaceted roles of DNA methylation in neoplastic transformation, from tumor suppressors to EMT and metastasis. Genes. 2020;**11**(8):922

[23] Yi JM. DNA methylation change profiling of colorectal disease: Screening towards clinical use. Life. 2021;**11**(5):412

[24] Fatemi N, Tierling S, Es HA, Varkiani M, Mojarad EN, Aghdaei HA. i in. DNA methylation biomarkers in colorectal cancer: Clinical applications for precision medicine. International Journal of Cancer. 2022;**2022**:29. https://onlinelibrary.wiley.com/doi/abs/10.1002/ijc.34186

[25] Cervena K, Siskova A, Buchler T, Vodicka P, Vymetalkova V. Methylation-based therapies for colorectal Cancer. Cells. 2020;**9**(6):1540

[26] Zhao G, Liu X, Liu Y, Li H, Ma Y, Li S. Aberrant DNA methylation of SEPT9 and SDC2 in stool specimens as an integrated biomarker for colorectal cancer early detection. Frontiers in Genetics. 2020;**2020**:11. https://www.frontiersin.org/articles/10.3389/fgene.2020.00643

[27] Zhao G, Liu X, Liu Y, Ma Y, Yang J, Li H, i in. Methylated SFRP2 and SDC2 in stool specimens for colorectal Cancer early detection: A cost-effective strategy for Chinese population. Journal of Cancer 5 marzec 2021;12(9):2665-2672.

[28] Xu T, Ding H, Chen J, Lei J, Zhao M, Ji B, et al. Research Progress of DNA methylation in endometrial Cancer. Biomolecules. 2022;**12**(7):938

[29] Pallarés J, Velasco A, Eritja N, Santacana M, Dolcet X, Cuatrecasas M. Promoter hypermethylation and reduced expression of RASSF1A are frequent molecular alterations of endometrial carcinoma. Modified Pathology. 2008;**21**(6):691-699

[30] Multinu F, Chen J, Madison JD, Torres M, Casarin J, Visscher D, et al. Analysis of DNA methylation in

endometrial biopsies to predict risk of endometrial cancer. Gynecologic Oncology. 2020;**156**(3):682-688

[31] Pabalan N, Kunjantarachot A, Ruangpratheep C, Jarjanazi H, Christofolini DM, Barbosa CP. Potential of RASSF1A promoter methylation as biomarker for endometrial cancer: A systematic review and meta-analysis. Gynecologic Oncology. 2017;**146**(3):603-608

[32] Ebata T, Yamashita S, Takeshima H, Yoshida H, Kawata Y, Kino N, et al. DNA methylation marker to estimate ovarian cancer cell fraction. Medical Oncology. 2022;**39**(5):78

[33] Ishihara H, Yamashita S, Amano R, Kimura K, Hirakawa K, Ueda T, et al. Pancreatic Cancer cell fraction estimation in a DNA sample. OCL. 2018;**95**(6):370-379

[34] Hentze JL, Høgdall CK, Høgdall EV. Methylation and ovarian cancer: Can DNA methylation be of diagnostic use? Molecular Clinical Oncology. 2019;**10**(3):323-330

[35] Miller B, Petrykowska H, Elnitski L. Assessing ZNF154 methylation in patient plasma as a multicancer marker in liquid biopsies from colon, liver, ovarian and pancreatic cancer patients. Scientific Reports. 2021;**11**:221

[36] Kubiliūtė R, Žukauskaitė K, Žalimas A, Ulys A, Sabaliauskaitė R, Bakavičius A, et al. Clinical significance of novel DNA methylation biomarkers for renal clear cell carcinoma. Journal of Cancer Research Clinical Oncology. 2022;**148**(2):361-375

[37] Locke WJ, Guanzon D, Ma C, Liew YJ, Duesing KR, Fung KYC, et al. DNA methylation Cancer biomarkers: Translation to the clinic. Frontiers in Genetics. 2019;**2019**:10. https://www.frontiersin.org/articles/10.3389/fgene.2019.01150

[38] Hibler E, Huang L, Andrade J, Spring B. Impact of a diet and activity health promotion intervention on regional patterns of DNA methylation. Clinical. Epigenetics. 2019;**11**(1):133

[39] Ullah MF, Usmani S, Shah A, Abuduhier FM. Dietary molecules and experimental evidence of epigenetic influence in cancer chemoprevention: An insight. Seminars in Cancer Biology. 2022;**83**:319-334

[40] Boughanem H, Kompella P, Tinahones FJ, Macias-Gonzalez M. An overview of vitamins as epidrugs for colorectal cancer prevention. Nutrition Reviews. 2022;**26**:nuac065

[41] Williams EA. Folate, colorectal cancer and the involvement of DNA methylation. Proceedings of the Nutrition Society. 2012;**71**(4):592-597

[42] Yousefi PD, Suderman M, Langdon R, Whitehurst O, Davey Smith G, Relton CL. DNA methylation-based predictors of health: Applications and statistical considerations. Natural Review in Genetics. 2022;**23**(6):369-383

[43] Ao C, Gao L, Yu L. Research progress in predicting DNA methylation modifications and the relation with human diseases. Current Medicinal Chemistry. 2022;**29**(5):822-836

Chapter 4

Environmental Epigenetics and Obesity

Ivonne Ramirez-Diaz, Sagrario Lobato, Andrea Perez-Gonzalez, Alejandra Morales and Karla Rubio

Abstract

In recent years, increasing interest on the effects of dietary components on epigenetic processes and, consequently, on the regulation of gene expression and metabolic responses has led clinical efforts worldwide to approach obesity. When inadequate, food consumption leads to chronic and non-communicable diseases (CNCD) including obesity. Among the dynamic changes in cellular responses by nutritional interventions, epigenetic control represents a master regulator underlying both positive and negative effects of diet on body mass, including DNA methylation, histone post-translational modifications and microRNA expression signatures. Indeed, mechanistical studies of the relationship between environment, diet and differential epigenetic landscapes are gaining attention on functional pathways involved in cell growth, DNA-repair, lipogenesis, senescence, inflammation, tumor suppression, apoptosis and oncogenesis. Being the dynamic interplay between epigenetics and obesity so complex, moreover considering a detrimental environment context, this chapter will discuss the state-of-the-art evidence showing the pollution impact on the different epigenetic mechanisms regulating an obese phenotype, and how these molecular events determine the organic interplay upon metabolic alterations, and finally we will introduce recent epidrugs and biocompounds of therapeutic interests due to their potential to modulate and even revert obesity-inducing epigenetic mechanisms.

Keywords: toxicants, chromatin, ncRNAs, DNA methylation, endocrine disruptor

1. Introduction

Obesity epidemics has become pandemic in the last decade, placing a significant burden on the global health system. Although the heritability of the disease is high, all identified genetic variants associated with obesity represent a very small percentage of the phenotypic variation. Thus, the origins of obesity cannot be explained exclusively by genetic factors. In recent years, epigenetic studies have provided valuable information for a deeper understanding of the significant increase in global rates. Specific factors have related obesity to fundamental epigenetic changes, such as intrauterine environment, nutrition, circadian rhythms, psychosocial inputs, lifestyles, and a set of environmental stimuli. Therefore, health is itself the result of the interaction of the social (agriculture, industry, and energy production, use and management of water and waste, urbanization, income distribution, public services), the physical–chemical (soil, air, water, food, pathogens, climate, pollutants) and the biological environment (flora, fauna, habitats including reservoirs and vectors). Obesity-related environmental pollutants function in the body as endocrine disruptors, altering body weight, adipose tissue expansion, circulating lipid profiles, and adipogenesis through epigenetic mechanisms. To mention some introductory examples, it has been reported that widely diffused toxins, mainly bisphenol A, phthalates and pesticides, can promote obesity in children and adults, by acting on the differentiation pathway that unites multipotent stromal stem cells with mature adipocytes, modulating epigenetic factors and influencing in a series of mechanisms that ultimately lead to altered eating habits, increased adipocyte formation and fat storage.

2. A "Globese" reality

Currently it exists a debate that attempts to redefine the concept of obesity (**Figure 1**). Until now, obesity was considered as the Body Mass Index (BMI) equal

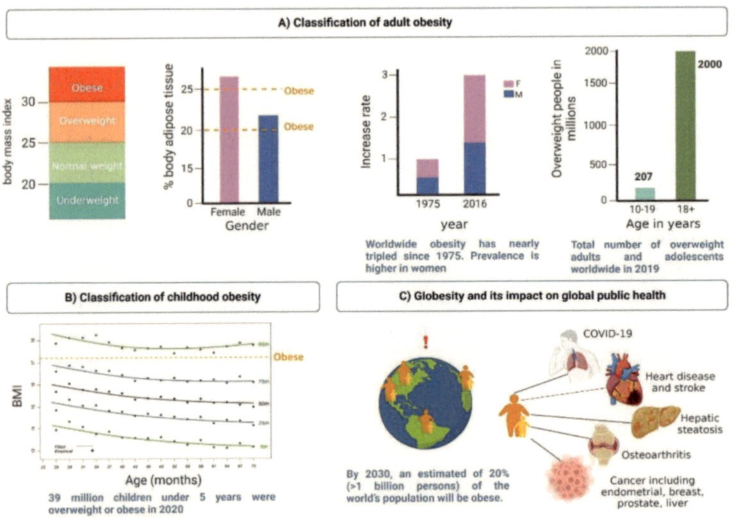

Figure 1.
Globesity. (A) Adult obesity is the body mass index (BMI) equal to or greater than 30 kg/m² and based on the percentage of body adipose tissue, when it exceeds 20% in men and 25% in women [1–3]. (B) Childhood obesity (under 5 years of age) is the BMI percentile equal to or greater than 95 according to the World Health Organization (WHO)'s growth standards [4–7]. (C) "Globesity" is a concept that refers to the obesity pandemic and an invitation by the WHO to create social awareness about its impact on global public health [8], since its health impacts range from an enhanced risk of premature death to significant chronic illnesses that diminish the overall quality of life. After smoking, it is the most important predisposing factor for the global burden of disease, contributing to the morbidity and mortality of COVID-19, type 2 diabetes mellitus, cardiovascular diseases, non-alcoholic fatty liver disease, hepatic steatosis, rheumatoid osteoarthritis, gout, musculoskeletal injuries, metabolic syndrome; as well as cancer types such as breast, endometrial, esophageal, colon, kidney and prostate; chronic respiratory diseases, sleep apnoea, fertility problems in both sexes, chronic kidney disease, neurodegenerative disorders, and maternal complications, according to the world obesity atlas** [9–12]. "Globesity" is higher in the afro-descendant and Hispanic population, and the increase in prevalence is more prominent in women [13, 14]. *https://www.who.int/toolkits/child-growth-standards/standards/body-mass-index-for-age-bmi-for-age, **https://www.worldobesity.org/resources/resource-library/world-obesity-atlas-2022).*

to or greater than 30 kg/m2 [1, 2]. However, due to the lack of tissue specificity, there is another definition of obesity related to the percentage of adipose tissue in the body, established at a maximum limit of 20% for men and 25% for women [3]. In the case of childhood obesity, the criteria are not so specific, but it is still considered as a basis that the body weight exceeds the relationship with the height of the child or adolescent [4, 5]. In this regard, the Center for Disease Control and Prevention of the National Public Health Agency of the United States recommends BMI charts for people from 2 to 20 years of age, which are color-coded based on the BMI percentile: 5 (red), 5–85 (green), 85–95 (yellow) and 95 (red). Thus, overweight is diagnosed between percentiles 85 to 95, and obesity with percentiles above 95 [6, 7].

Obesity is currently one of the most alarming pathological conditions due to its exacerbated prevalence (**Figure 1**) [15, 16]. Due to this reason the World Health Organization has proposed the term "globesity" [8] to raise awareness on the damage to global health caused by this disease in the last 50 years [17]. Prevalence is higher in the Afro-descendant and Hispanic population, and the increase in prevalence is more prominent in women [13, 14]. Obesity reduces both quality of life and life expectancy. After smoking, it is the most important predisposing factor for the global burden of disease, contributing to the morbidity and mortality of COVID-19, type 2 diabetes mellitus, cardiovascular diseases, non-alcoholic fatty liver disease, hepatic steatosis, rheumatoid osteoarthritis, gout, musculoskeletal injuries, metabolic syndrome; as well as cancer types such as breast, endometrial, esophageal, colon, kidney and prostate; chronic respiratory diseases, sleep apnoea, fertility problems in both sexes, chronic kidney disease, neurodegenerative disorders, and maternal complications [9–12] (**Figure 2**).

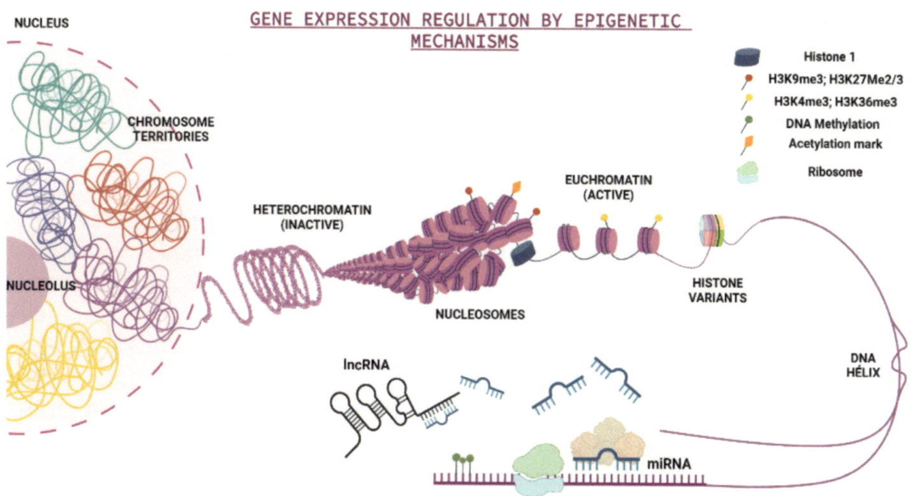

Figure 2.
Epigenetic mechanisms and topological structure of the nucleus. Gene expression can be regulated by non-genetic heritable genomic modifications without alterations in DNA sequence. Epigenetic mechanisms include DNA methylation, histone modifications and non-coding RNAs (ncRNAs), all regulating gene activity at the transcriptional, post-transcriptional, translational, and post-translational level. Moreover, lncRNAs participate in numerous biological activities, including cell cycle control, cytoplasmic and nuclear trafficking, splicing, transcription, translation, imprinting, epigenetic regulation, and more recently shown, in the arrangement of functionally different nuclear sub-compartments such as nucleolus and chromosomal territories. Thus, specific 3D nuclear topology or architecture is strongly related with normal cell functions.

3. Environmental pollution and its impact on epigenetics

Environmental change on our planet is a natural and ancient phenomenon that has occurred for over 4 billion years. However, the human being has left an ecological footprint affecting planet Earth, especially from the 20th century onwards [18, 19]. Environmental pollution is one of the most serious problems facing biodiversity, ecosystems and human health. It is widely defined as the introduction of toxins that alter the physical and biological components of the environment. Pollutants can be liquids, solids and harmful gases produced in concentrations that exceed the levels allowed by the World Health Organization (WHO), reducing the quality of life and the environment [20]. Specifically, air pollution is defined as the release of harmful particles into the air by one or more harmful gases. The six main air pollutants are: airborne fine particulate matter 10 microns or less in diameter (PM10) and 2.5 microns or less in diameter (PM2.5); ozone, carbon monoxide, lead, nitrogen dioxide and sulfur dioxide [20, 21]. Two main types of air pollution are considered: outdoor pollution as the pollution of the ambient air, and indoor pollution as the pollution generated by domestic combustion [20].

According to the United Nations, Educational, Scientific and Cultural Organization (2021), one in nine people in the world consumes water from unimproved and unsafe sources. 90% of wastewater in countries with developing economies is discharged directly into untreated water bodies. Moreover, wastewater is reused in agriculture, while important for livelihoods, it is associated with serious health risks (International Initiative on Water Quality, IIWQ, UNESCO, 2021). 99% of the world population breathes air that exceeds the maximum limits for air pollutant levels. Currently, more than 6000 cities in 117 countries monitor air quality, but their inhabitants still breathe unhealthy levels of fine particulate matter and nitrogen dioxide, with people in low- and middle-income countries experiencing the highest exposures (The report and WHO air quality database 2022). Additionally, the WHO declared airborne fine particulate matter the number one global environmental health concern in October 2021 [22]. Moreover, soil contamination by heavy metals has become a global health problem. According to The United States Environmental Protection Agency (2021), the sources of pollution in indoor environments are enlisted in **Table 1** [23].

In recent years, scientific evidence linking air pollution with obesity has been published by international groups, proposing several key biological pathways acting as a functional bridge [24, 27]. It is known that the exposome (exposure to environmental compounds and factors such as stress, habits, diet, exercise, lifestyle) can affect molecular pathways and cellular processes that increase the susceptibility to develop several diseases including obesity, because of abnormal epigenetic modifications [21, 28]. Early in the forties, Conrad Waddington first introduced the term "epigenetics" to describe environment-gene interactions that could not be explained by traditional Mendelian genetics [29]. Nowadays, epigenetics is defined as the study of non-genetic heritable genomic modifications that regulate gene expression without alterations in DNA sequence. Epigenetic mechanisms include DNA methylation, histone modifications and non-coding RNAs (ncRNAs), all regulating gene activity at the transcriptional, post-transcriptional, translational, and post-translational level (**Figure 2**) [21, 30].

DNA methylation is the most studied and characterized epigenetic mechanism, using as effectors the DNA methyltransferases (DNMT1, DNMT3A, DNMT3B, and DNMT3). These enzymes transfer a methyl group from S-adenosylmethionine (SAM) onto the C5 position of the cytosine to form 5-methylcytosine (5mC), the major

	Air pollution	Water pollution	Soil pollution
DEFINITION	Release of harmful particles into the air by one or more harmful gases. Two main types are considered: outdoor pollution as the pollution of the ambient air, and indoor pollution as the pollution generated by domestic combustion [20].	Deterioration of water quality worldwide due to the acceleration of industrialization.	Accumulation of organic and inorganic contaminants leading to unbalanced availability of nutrients to plants, changes in the abundance and structure of the soil microbial community, degradation of the soil ecosystem, and contamination of groundwater; further affecting the quality and safety of crops and human health.
SOURCES OF EXPOSURE	Indoor air [23]:	Contamination of waterbodies by waste from the agricultural, industrial, mining and energy sectors.	Agricultural production, the quality of cultivated land and environmental hygiene habits of the population.
	Combustion processes: heating, cooking food, smoking and fireplaces.		
	Cleaning products, paints, insecticides, and other commonly used products.		
	Building materials, either through degrading materials (i.e. asbestos fibers) or new materials (i.e. chemical off-gassing from pressed wood products).		
	Other substances such as radon, mold, and pet dander.		
	Outdoor air [23]:		
	Contaminants entering indoors through open doors and windows, ventilation systems, and cracks in structures.		
	Power plants, refineries, petrochemicals, chemical and fertilizer industries, metallurgical plants, and municipal incineration.		
	Mobile sources: automobiles, railways, airplanes, and other types of vehicles.		
	Natural sources: wildfires, volcanic eruption, dust storms, and agricultural burns.		

	Air pollution	Water pollution	Soil pollution
COMPOUNDS	The six main air pollutants are: airborne fine particulate matter 10 microns or less (PM_{10}) and 2.5 microns or less in diameter ($PM_{2.5}$); ozone, carbon monoxide, lead, nitrogen dioxide and sulfur dioxide [20, 21].	Agrochemicals, pathogens, nutrients and metals: Cr, Al, Ba, Cu, Mo, Ni, Pb, Se, As and Zn.	Agrochemicals, pathogens and heavy metals such as Cd, Pb, Cr, Zn, and Cu.
RISK FACTORS	Industrialization, population explosion, and fossil fuel economy [24]. The most prominent risk factor is long-term exposure, but there is no operational definition in terms of distance and time established by regulatory bodies, such as the World Health Organization. Regarding exposure to $PM_{2.5}$, several authors agree on the distance between the residence and the nearest highway or main road. A study reports the distance as 7 km between the residence with the automatic station for monitoring air quality from which the verified data was obtained, other studies coincide with the sites measurement methods mentioned above, establishing the exposure distance of 30 km, 10 km, 1.2 km or 1 km. 99% of the world population breathes air that exceeds the maximum limits for air pollutant levels. Currently, more than 6000 cities in 117 countries monitor air quality, but their inhabitants still breathe unhealthy levels of fine particulate matter and nitrogen dioxide, with people in low- and middle-income countries experiencing the highest exposures (WHO air quality database, 2022). Additionally, the World Health Organization declared airborne fine particulate matter the number one global environmental health concern in October 2021 [22].	Toxins with a greatest risk are metals and can be ingested by humans in crops that were irrigated with contaminated water. 1/9 people in the world consumes water from unimproved and unsafe sources. 90% of wastewater in countries with developing economies is discharged directly into untreated water bodies. Moreover, wastewater is reused in agriculture, while important for livelihoods, it is associated with serious health risks (International Initiative on Water Quality, IIWQ, UNESCO, 2021). Heavy metals in drinking water can induce genetic and epigenetic modifications that affect gene expression in growth control genes such as DNA-repair, tumor suppressor, apoptotic or oncogenes [25]. Susceptibility depends on surface and ground water use for drinking, cooking, washing, among others as part of their daily routine [26].	Concentrations of heavy metals in urban soils vary significantly depending on the city, the type of land used, the population density and the volume of traffic.

Table 1.
Sources of pollution in air, water and soil.

form of DNA modification. 5mC is present mainly in CpG sites on gene promoters, with relevant roles in development and disease, and conventionally associated with gene silencing or transcriptional repression [31]. The methylation repressive effect is caused by the attraction or the repulsion of certain DNA-binding proteins. Repressor complexes have been observed to be recruited to methylated promoter regions by a class of proteins known as methyl-CpG binding domain proteins (MBDs), which are attracted to and bind DNA-containing methylation CpG dinucleotides [32]. In addition, exposure to toxicants found in the environment may promote differential DNA methylated regions (DMR), potentially contributing to the obese phenotype [33, 34].

Histone modification is another well-known epigenetic mechanism [35]. Since DNA is packaged in the form of chromatin, its basic unit the nucleosome contains 147 bp of DNA wrapped around an octamer of histone proteins formed of two dimers of H2A and H2B, and a tetramer of H3 and H4 proteins, and an H1 as linker of adjacent nucleosomes [36]. Histone tails (N- and C- terminal) can be post-translationally modified by methylation, acetylation, phosphorylation, ubiquitylation, ADP-ribosylation, citrullination, sumoylation, carbonylation and proline isomerization, among others, thus, directly regulating chromatin structure [36, 37]. Methylation and acetylation are the most studied mechanisms occurring on histone tails. Methylation marks can cause either transcriptional repression (i.e., H3K9me3 and H3K27me2/3) or activation (i.e., H3K4me3, H3K36me3 and H3K79me3) and occurs mainly in arginine and lysine residues catalyzed by histone methyltransferases (HMTs). Reactions can be reverted by several histone demethylases (HDMs) such as KDM1 or AOF [38]. Like methylation, histone acetylation regulates chromatin remodeling (euchromatin and heterochromatin) and affects gene expression by dynamic changes through the action of histone acetyltransferase (HATs), also known as "writers", and histone deacetyltransferases (HDACs), also called "erasers" [26474904]. HDACs are divided into four classes, class I (HDACs 1, 2, 3 and 8), II (HDACs 4, 5, 6, 7, 9 and 10), III or sirtuins, and IV with HDAC11 as unique member. Each group harbors specific functions, i.e. HDACs class I play a role in adipocyte differentiation as well as in establishing the metabolic characteristics of these cells [39, 40].

Around 98% of the transcriptome consists of non-coding RNAs (ncRNAs) that cannot be translated into proteins, so they contribute to the physiological complexity of mammals. Non-coding RNAs are classified as 1) small-ncRNAs (20–30 nucleotides length) and 2) long-ncRNAs (lncRNAs, >200 nucleotides length), which are subclassified based on their biosynthesis and effector proteins. The main functions of small-ncRNAs are related to gene expression at transcriptional, post-transcriptional and translational levels. Moreover, they can be grouped in small interfering RNAs (siRNAs), Piwi-interacting RNAs (piRNAs) and microRNAs (miRNAs) as the most study biotypes [41]. Meanwhile, lncRNAs participate in numerous biological activities, including cell cycle control, cytoplasmic and nuclear trafficking, splicing, transcription, translation, imprinting, epigenetic regulation, and more recently shown, in the arrangement functionally different nuclear sub-compartments and their impact on nuclear topology or architecture [42, 43]. Hence, several human diseases like obesity are linked to alteration of ncRNA biosynthesis and function.

3.1 Pollution impact on DNA methylation

Environmental epigenetics explains the biological pathways that are altered by environmental factors that modify epigenetic mechanisms [44]. These changes include as main mechanisms DNA methylation, histone post-translational modifications, ribonucleic acid-based mechanisms, and chromatin remodeling [45]. DNA methylation is an epigenetic mechanism that consists of the covalent transfer of a methyl group to the C5 position of the DNA cytosine ring by DNA methyltransferases (DNMT1, DNMT3A, DNMT3B, and DNMT3) [46, 47]. In mammals, methylation occurs primarily at the cytosine residues of cytosine and guanine dinucleotides (CG), but CG dinucleotides within promoters are usually free of methylation [46, 48].

Furthermore, the functions of DNA methylation differ depending on its location. In promoter regions, it confers gene repression, whereas, throughout the gene body, it is associated with transcriptional activation. Methylation levels also depend on genetic sequence and DNA-binding factors. Therefore, there is no general rule that can be applied to all biological situations, reflecting the high complexity of DNA methylation-dependent regulatory pathways [46, 49]. During all stages of human life, DNMTs can add or remove methyl groups by ten-eleven translocation enzymes (TETs), and because the methylome (global genome methylation state) operates at the interface between the genome and the environment, it can be modified in response to environmental stimuli, such as exercise, diet, smoking, or pollutants [48]. Altered patterns of DNA methylation have been associated with obesity and other chronic degenerative pathologies [50].

A large percentage of individuals worldwide are exposed to high arsenic (As) concentrations (10 ppb) in drinking water [51]. A revealing study was made in mononuclear cells of peripheral blood from individuals exposed to well-water. Araihazar's concentrations (0.1–960 μg/L), a region constituted by 10,000 wells near to Dhaka, Bangladesh. Forty participants were divided in two groups: low water concentration of As (median 55 μg/L, range 50–81 μg/L) and high concentration (median: 216 μg/L; range:150–500 μg/L) [52]. Interestingly, they found that As exposure influenced histone marks in a gender specific manner. H3K18ac and H3K27ac levels were higher in males than in females; in contrast, H3K27me3 and H3K4me3 increased in females. A probable explanation is that As has an endocrine disruptor role by mediating the estrogenic receptor (*ER*) expression. Also, these histone marks are known as estrogenic-sensitive which may have a relationship with the As effect. However, they did not fully characterize specific genes and pathways that were modified by As chronic exposure [52].

On the other side, chlorination by-products of drinking water (i.e. triethyltin, chloroform, dichloroacetic acid, trichloroacetic acid, bromodichloromethane, chlorodibromomethane and bromoform) have shown carcinogenic properties, modifying the methylation profile in liver and kidney in mice [53]. In this line of ideas, a human normal hepatic cell line treated with 0.1–0.9 mM trichloroacetic acid showed lower expression levels of Histone deacetylases (HDACs, mainly *HDAC2* and *HDAC3*) after 24 h compared to control treated cells. Intracellular H3K9ac levels increased and PCAF was maintained at low levels, resembling a DNA methyltransferase inhibitor (5-aza-dC) effect. Remarkably, TCA effect in HDACs was reversed after 72 h, which suggests that TCA long exposure cause DNA hypomethylation in promoter regions of *HDACs* genes, therefore activating their expression. As discussed by the authors, *HDACs* and H3K9ac levels could represent early epigenetic biomarkers useful for toxicity evaluation to prevent disease development upon TCA exposure [54]. Nevertheless, it is necessary to elucidate the relationship between early and long TCA effect on histone acetylation and DNA methylation.

In addition, a cyanobacterial toxin called cylindrospermopsin (CYN) is uptaken for humans while drinking water and through contaminated food by bioaccumulation. Recent studies characterizing the transcriptomic profile of the colorectal adenocarcinoma cell line Caco-2 showed chromatin remodeling events after CYN exposure. Precisely, 2911 differentially expressed genes appeared between CYN treated and control cells. Among these genes, authors identified enzymes forming the RNA polymerase II complex (*POLR2D*, *POLR2L*, *POLR3E* and *POLR1C*), transcription co-activation factors (*MED6*, *MED10*, and *MED21*), enzymes involved in RNA maturation, acetyl transferases (*MYST1*, *KAT5*) and methyl transferases (*EHMT2*). Also, they confirmed a CYN-increased gene expression for *POLR2D*, *POLR2L*, *MED6*, *DDX20*, *KAT5*, *MYST1* AND *EHMT2*, as well as a differential level of proteins KAT5, MYST1, and EHMT2. Interestingly, different histone marks were modified in the same conditions, for example acetyl-histone H2A (Lys5), methylLys4 and Lys9 on histone H3 reflecting the activity of EHMT2, and dimethyl-Lys4 on histone H3. In this manner, important determinations were made in the context of environmental epigenome reprogramming in CYN-treated CaCo cells [55].

3.2 Pollution impact on histone post-translational modifications

Covalent histone modifications have a significant role during gene expression, inducing open chromatin for active transcription, and condensate chromatin for inactive transcription [54]. In recent years, the term obesogens has emerged referring to compounds that interrupt lipid homeostasis and promote adipogenesis. Several obesogens like heavy metals, solvents, pesticides, PCBs, organic phosphates, phthalates, organotin and diethylstilbestrol (DES) remodel histone marks associated to inflammatory and stress responses [56].

Systemic inflammation is widely linked to air pollution and obesity [57, 58]. However, the mechanism that connect both processes is poorly understood. Monocytes from obese individuals have shown elevated levels of IL6 and H3K9/H3K18 acetylation, associated to EP300, transcription factors and RNA Polymerase II recruitment to *IL6* promoter regions. EP300 silencing, and inhibition of histone acetyltransferase attenuated this effect. In this sense, a cohort study with individuals exposed for long periods to particulate matter (PM) (aerodynamic diameter ≤ 2.5, 2.5–10, and ≤ 10 μm) and gaseous air pollutants confirmed the increased *IL6* levels compared to short exposures. These studies provided crucial highlights about histone code changes in response to environmental pollutants during inflammatory events [59].

AMP-activated protein kinase (AMPK) is a known sensor of cellular energy homeostasis. AMPK phosphorylates targets involved in lipid homeostasis, mitochondrial biogenesis, and glycolysis, upon its stimulation above low energy conditions [60]. Recently, it was reported the genome-wide transcriptional profile and chromatin landscape of pancreatic islets from mice fed high-fat diet (HFD) treated with O304, a pan-AMPK activator. O304 treatment on HFD abrogated *Aldh1a3* expression, a β-cell stress marker, and *Ins1* mRNA levels, whereas it restored the glucose transporter *Slc2a2* expression, a target of impaired glucose response and insulin secretion. Screening of histones marks in O304-treated pancreatic islets indicated an ~58% active chromatin marks and preserved ~27% of H3K27Ac, compared to regions in HFD islets. Particularly, HFD-O304 cells diminished H3K27Ac in the *Aldh1a3* promoter as well as two distant upstream regions (~ 70 kb and ~ 114 kb) [60]. Tetrabromobisphenol A (TBBPA)-exposed human adenocarcinoma hepatic cells and Cr (IV)-treated neutrophils increase ROS production which mediate AMPK activation and promote cellular proliferation and apoptosis [61, 62]. In the case of transcriptomic analysis in TBBPA-exposed adenocarcinoma liver cells revealed differential expression in growth factors *FGF17* and *EFN5A*, Ras signaling pathway activators, involved in cell detoxification mechanism, lipid and vitamin metabolic rate regulation [61]. On the other hand, Cr (IV)-exposed neutrophils showed the reduction on myeloperoxidase (*MPO*) and *H3* expression, which inhibited neutrophil extracellular traps (NETs) formation, a deconcentrated chromatin scaffolds complex. Interestingly, metformin treatment in Cr (IV)-exposed neutrophils attenuated heavy metal effect on ROS levels through nuclear factor (erythroid-derived 2)-like 2 (*NFE2L2*) induction, a key transcription factor of antioxidant genes, apoptosis, NETs formation, and protein imbalance [62]. Taken together, O304, a pan-AMPK activator and metformin treatment may benefit restoration of glucose homeostasis and insulin sensitivity in obesity patients after chromatin remodeling by toxicants exposure.

3.3 Pollution impact on ncRNAs

Implementation and integration of omics data, such as epigenomics, transcriptomics, proteomics, and metabolomics, is increasingly used to detect early and subtle molecular responses to environmental compounds. Thus, multi-omics profiles such as, circulating miRNAs, blood DNA methylation marks, gene expression, proteins, urine and serum metabolites, may provide a broad perspective on all cellular activities [63]. Regarding to epigenomics, DNA methylation may be more useful for epidemiological research that compare certain individuals based on a single measurement, whereas ncRNAs profiles may provide more valid information for studies that analyze individual trajectories across time [63]. Despite the importance of epigenetic modifications that occur directly on the DNA strand, noncoding RNAs (ncRNAs) represent a wide and well-orchestrated regulatory mechanism of gene expression. Most (98.5%) of the eukaryotic genome is transcribed into ncRNAs, including microRNAs (21–25 nucleotides in length) and long noncoding RNAs (lncRNA, > 200 nucleotides in length, or lacking an open reading frame of >100 amino acids) [64, 65]. Although miRNAs are supposed to act mainly in the cytosol by inhibiting translation, recent research has shown their location, activity and relevance in the nucleus of cells as direct regulators of cell phenotype [66–69]. MicroRNAs guide Argonaut (Ago) proteins to specific target mRNAs leading to their destabilization or translational repression. The mature miRNA acts to guide not only Ago but also the RNA-induced silencing complex (RISC) ribonucleoprotein complex. On the other hand, lncRNAs are important regulators of different biological processes in the nucleus, such as providing a framework for the assembly of defined chromatin structures at specific loci, thereby modulating gene expression, centromere function, and silencing of DNA repetitive elements [70].

Non-coding RNAs are RNA molecules that are not translated into proteins but that importantly modulate gene expression in specialized cellular processes such as adipogenesis [71]. Several works provide evidence regarding the potential role of microRNAs in metabolic disorders, specifically type 2 diabetes and obesity [71]. For instance, Kunej et al. identified 221 miRNAs to be dysregulated in distinct species. Among them 14 miRNAs, including *let-7a, let-7b*, *let-7c, let-7e, let-7f*, *mir-103*, *mir-10b*, *mir-125a*, *mir-125b*, *mir-143*, *mir-23a*, *mir-23b*, *mir-26a*, and *mir-99b* directly impact fat accumulation in cattle, rats, mice and humans [72]. In humans, diet and lifestyle directly influence the expression of microRNAs such as *miR-17/20/93*, *miR-21/590-5*p, *miR-200b/c*, *miR-221/222*, *let-7/miR-98* and *miR-203* families are the most dysregulated in this setting [73].

ncRNAs have also the potential to mediate the cellular response to environmental toxicants. Environmental stressors or environmental obesogens [74] may be strongly related with obese phenotype by aberrant miRNAs expression [75]. For instance, phthalates, a class of plasticizers, are widely employed in a variety of everyday items. Recent research implicating butyl benzyl phthalate (BBP) as an obesogen has increased public health concerns [76]. Meruvu et al. demonstrated that the expression of *miR-34a-5p* and its target genes, *NAMPT* and *SIRT1*, is perturbed when developing 3 T3-L1 cells are exposed to varying concentrations of BBP without external adipogenic stimuli [77]. Exposure to BBP increased the expression levels of *miR-34a-5p*, resulting in a reduction in *NAMPT* and *SIRT1* and a subsequent rise in adipogenesis [77]. Moreover, Mcllwraith et al. recently demonstrated that *miR-708-5p* mediates the effects of BPA in hypothalamic cells through the reduction of neuronatin levels and the increase in orexigenic Neuropeptide (Npy) [78]. Contrarily, Rahmani et al. showed revealed that the expression of *miR-375*, *miR-676*, *miR-126-a*, and *miR-340-5p* was significantly disrupted by BPA, resulting in aberrant β-cell metabolism and diabetes [79]. More interestingly, the interplay between environment and gene expression seems to be a coordinated action of different ncRNA biotypes, at least lncRNAs and miRNAs. In the case of chronic exposure of human primary adipocytes from Caucasian females to BPA, but also other structural analogs like bisphenol F (BPF) or S (BPS), long intergenic non-coding RNAs (upregulated *LINC01140* and *LINC01088*, downregulated *LINC01048*), small nucleolar RNAs and miRNAs (*MIR-4655-3p,* upregulated *MIR-30c*, downregulated *MIR-136*) were differentially expressed in a coordinated manner with conventional genes related with adipocyte differentiation (*LPL*, *PLIN1/4*, *ADIPOQ* and *FABP4*) [80].

Not only the direct cell-cell contact is used endogenously to induce tissue-specific responses upon harmful stimuli. The release of extracellular membrane vesicles (EVs) has been on the scope during the recent years as a potential modifiable scenario in terms of pathological tissue reprogramming. DNA, RNA, protein and lipids-containing EVs are released in all body fluids, and importantly, their cargo miRNAs dysregulation can induce phenotypic changes on a distant target tissue. EVs and EV-associated miRNAs (EV-miRNAs), such as *let-7c-5p*, *miR-106a-5p*, *miR-143-3p*, *miR-185-5p*, *miR-218-5p*, *miR-331-3p*, *miR-642-5p*, *miR-652-3p* and *miR-99b-5p* have been considered as mediators of the detrimental effects of PM10 exposure, since their levels are reduced in EVs after short-term exposure and correlate with elevated fibrinogen levels and subsequent cardiac injury [81].

3.4 Impact on genome 3D structure

Recently developed Chromosome Conformation Capture (3C) technology, based on Hi-C, ChIA-PET, and Hi-C capture approaches, have revealed important hints about the role of chromosomal organization and compaction for transcription mechanisms. Precisely large, gene-poor chromosomes are frequently located in the nuclear periphery, whereas small, gene-rich chromosomes are located at the internal side of the nucleus. Chromosomes are divided into A compartments with active chromosome domains, and B compartments with inactive chromosome domains. For example, lamin-associated domains (LADs). These compartments have several topologically associated domains (TADs). TADs are multiple regulatory loops that include DNA sequences exhibiting significantly higher contact frequency with other DNA sequences, among distal enhancers and promoters, within a range from 500 kb to 1 Mb. TADs are mediated by CCCTC-binding factor (CTCF)/cohesin complex. In this case, altered CTCF occupancy is associated with several diseases due to aberrant chromosome looping between distal cis-regulatory elements and their target promoter(s), inducing altered gene expression [82].

Aberrant pancreas and adipose tissue function contributes with metabolic alterations triggering obesity-related diseases as diabetes, and insulin resistance [82, 83]. Gene expression signature in adipocytes is influenced by gene interactions with proximal and distal *cis* regulatory elements, which are mediated by 3D chromosome architecture maintenance in adipose tissue [82]. High-resolution maps of chromatin architecture of porcine livers under a high-fat diet-induced obesity demonstrated changes in genome 3D structure by similarities in Hi-C contact maps, compartmentalization strength, largely conserved TAD boundaries and intra-TAD contact intensities. Also, in terms of metabolic adaptation to excessive energy intake in pigs, even in comparison with variations in liver pig development. Among 160 genes exhibited significant expression changes, 126 was non-alcoholic fatty liver disease-related genes, including *ADIPOQ, CYP2E1, IL6, LEP, TNF*, and target genes of HNF4α and C/EBPα. In addition, it was observed hepatic morphology modifications, increase in lipid accumulation, and serum concentrations of five metabolic indicators (triglyceride (TG), low-density lipoprotein (LDL), high-density lipoprotein (HDL), cholesterol and glucose) which did not have significant changes in livers of pigs fed with HFD compared to pigs fed normal adult diets. These results suggested that pigs resist to 'diabetogenic' environments and increase their tolerance to chronic damage from obesity in the liver [84].

Recently, a research group studied metabolic adaptation to chronic obesity and NAFLD on male C57BL/6 mice with hypercaloric diet. The evaluation of transcriptomic profile showed 2066 genes overregulated in obese animals on the lipid-rich diet, and 1663 genes overregulated on the carbohydrate-rich diet. Interestingly, lipid-rich diet suppresses *de novo* lipogenesis in comparison with a carbohydrate-rich diet. In general, obese animals by lipid-rich diet showed differential expressed genes (DEG) involved in fatty acid metabolism such as type I acyl-CoA thioesterases (*Acot2, Acot3, Acot4, Acot5, Acot6*), and type II acyl-CoA thioesterases (*Acot7, Acot8, Acot9, Acot13*); beta oxidation of fatty acids like acyl-CoA dehydrogenases (*Acadm*, *Acads*, *Acadvl*), enoyl-CoA hydratase (*Ehhadh*), and hydroxyacyl-CoA dehydrogenase (*Hadh*), as well as mitochondrial carnitine-dependent lipid transporter, *Cpt1.* The evaluation of 3D chromosome architecture revealed that TADs and their boundaries are largely conserved in both diet groups. Furthermore, they identified lipid-rich diet enriched H3K27ac regions that corresponded to known consensus binding sequences for ETS (*ETS1, EHF*), bZIP (*FOSL2, JUN-AP1, and ATF3*), and C/EBP (*C/EBPA, C/EBPB, and C/EBPE*) transcription factors. C/EBP family participates in lipogenesis regulation. Motifs for the bZIP family transcription factors and nuclear receptors (*HNF1, HNF6*) involved in lipid and carbohydrate metabolism regulation for loci were found enriched upon carbohydrate-rich diet. On the other hand, promoter-chromatin interactions evaluation revealed 34,982 significant chromatin interactions in liver from animals on lipid-rich diet, and 37,185 from animals on carbohydrate-rich diet. Notably, they confirmed the association between promoter-interacting regions (PIRs) and H3K27ac enrichment with high gene expression. As a result, they established a promoter-interaction landscape in liver under different dietary regimens in response to obesity and metabolic stress [85].

Glyphosate is a widely used herbicide in agricultural activities. A study made in blood samples from workers exposed occupationally detected concentrations of 0.05–0.5 mM, even in people who was not directly exposed to this herbicide, with concentrations between 0.435 ± 0.167 μM. Peripheral blood mononuclear cells (PBMCs) treated with glyphosate and aminomethylphosphonic acid (AMPA), significantly increased gene expression of *DNMT1* and *DNMT3A*, involved in DNA methylation, as well as *HDAC3*-associated histone deacetylation [86].

Lately, a research group traced the offspring from pregnant mouse females (F0) exposed to BPA via intraperitoneal injection. Curiously, second generation of male and female mice without exposure (BPA-F2) increase their body weight, showed a large accumulation of visceral white adipose tissue, increased number of adipocytes and accumulation of lipid droplets, compared to control F2 animals, but these effects were completely lost in BPA-F7 animals. They performed *in vitro* fertilization confirming that median weight of males and females from the embryo transfer experiment was significantly higher than in controls. In this way, the authors confirmed that the overweight phenotype is due to epigenetic modifications in the germline rather than external factors. Then, they observed a dramatic redistribution of transcription factor (TFs) and increase of interactions in sperm exposed to BPA F1-F3, indicating the formation of new CTCF loops, adjacent to *Steap1* and *Steap4*, genes involved in obesity in humans. Also, among 1327 differential sites, 107 are conserved in germ lines through generations, whereas other sites disappear. 69 BPA-gained sites in distal intergenic regions or introns corresponded to motifs for several TFs such as *Znf143, Foxa1,* and nuclear hormone receptors, including *Esr1/2, AR, and Ppar.* In sperm from BPA-F3 males, 1428 hypomethylated regions and 648 hypermethylated regions were identified to persist in a transgenerational manner. Furthermore, these 69 enriched sites interacted with 610 gene promoters and enhancer regions, involved in obesity, corresponding with the enrichment of H3K27ac. These results suggested that BPA, directly or indirectly, promote the binding of CTCF and various transcription factors at genomic sites that are normally methylated. These gene promoters become activated at a higher frequency in exposed sperm BPA. This occurs mainly through interactions between proximal *Fto* enhancer and target promoters including *Rpgrip1l, Irx3, Irx5, Slc6a2, and Mmp2*, known to regulate body size and obesity by affecting appetite and food consumption. According to this meta-analysis, intron 8 of *Fto* gene seems to be an important element responsible for the transgenerational transmission of obesity after BPA exposure. Even though these findings, it is poorly understood if the exposure to other environmental toxicants can lead heritable or reversible modifications in epigenetic landscape that modify gene expression contributing with obesity or other pathologies.

4. Organic interplay between pollution and obesity

4.1 Digestive systems

Pollutants that enter the body through the oral route first affect the digestive tract and then, when they reach the systemic circulation, the rest of the body's homeostasis [87, 88]. The World Health Organization (2022) states that the contamination of food and drinking water by chemicals from the environment represents a threat to human health [89]. These chemical compounds belong to the groups of polycyclic aromatic hydrocarbons (PAHs), metals and metalloids, perfluorinated compounds (PFCs), persistent organic pollutants (POPs), and consumer products [90]. These compounds may disturb digestive tract and its normal functioning by affecting the normal microbiota (**Figure 3**).

The microbiota is the set of microorganisms hosted in a specific niche [91]. The human intestinal microbiota is made up of 1014 microorganisms that include bacteria, viruses, archaea and fungi. Intestinal bacteria in healthy individuals are made up of four phyla *Proteobacteria, Actinobacteria, Bacteroidetes* and *Firmicutes*. The

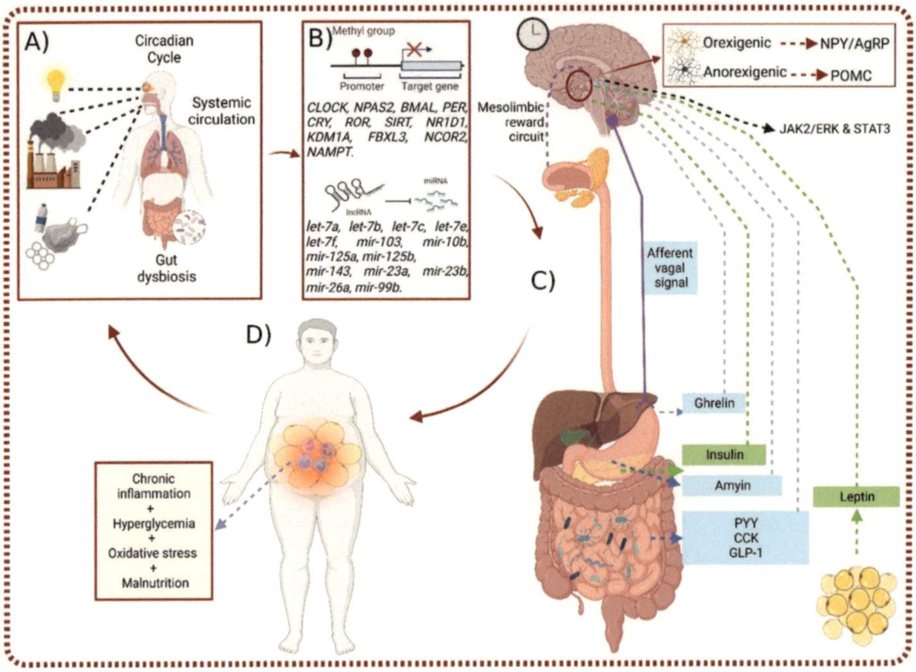

Figure 3.
Functional interaction between pollution, epigenetics and obesity. (A) Environmental contaminants such as PAHs, BPA, PFCs, POPs, light, particulate matter, among others, enter the body through respiratory, oral and ocular routes. (B) At the cellular level, pollutants cause changes in epigenetic landscapes that affect the expression profile of genes encoding proteins involved in energy homeostasis and obesity. (C) The above events alter the dynamic (blue) and static (green) signals of the hunger-satiation circuit in the neuroendocrine axis. In addition, they cause intestinal dysbiosis and, in cases of light pollution, alteration of the circadian cycle. (D) Endocrine disruption impacts metabolic pathways, mainly accelerating lipogenesis, which causes adipocyte hypertrophy and hyperplasia, increasing paracrine and endocrine activities driven by the adipose tissue. Therefore, pollution affects body systems through various epigenetic mechanisms that contribute to the global increase in obesity.

colon harbors a high density of the bacterial families *Prevotellaceae, Bacteroidaceae, Lachnospiraceae, Rikenellaceae*, and *Ruminococcaceae* [92, 93]. Intestinal microbiota has metabolic, immunological and trophic functions, fermenting indigestible dietary components, particularly undigested carbohydrates, to generate short-chain fatty acids [94, 95].

A direct relationship has been reported between exposure to fine particulate matter suspended in the air with intestinal dysbiosis and with obesity [96–98], which affects host adiposity through a comprehensive signaling pathways [99]. Environmental pollutants can influence the variety of microbiota, and their metabolites influence the activity of epigenetic enzymes. The mechanisms of action of environmental pollution also include the interaction between different niches of the human microbiota, such as the lung-gut axis [96, 100] (**Figure 3**).

4.2 Neural

Recently, artificial light at night has been suggested as an environmental factor that favors the appearance of obesity. Because most living things have developed circadian rhythms that are in sync with the daily cycle of light and dark, constant exposure to artificial light can disrupt the circadian rhythm and alter the secretion of various hormones, leading to disease metabolic, including obesity [101, 102]. The circadian rhythm is the internal manifestation of the solar day that allows adaptations to predictable environmental temporal changes. These ~24 h rhythms are controlled by molecular clockwork mechanisms in the hypothalamus that are reset daily to a precise 24 h by exposure to the light–dark cycle [103, 104].

Clock-controlled genes regulate circadian rhythms and the most important ones, as well as their products respectively, include: circadian locomotor kaput cycles (*CLOCK*) and its neuronal paralogous protein with PAS domain 2 (*NPAS2*), nuclear translocator of the brain aryl hydrocarbon receptor and muscle such as Arntl (*BMAL*), dot (*PER*), cryptochrome (*CRY*), retinoic acid-related orphan receptor (*ROR*), sirtuin (*SIRT*), nuclear receptor subfamily 1 group D member 1 (*NR1D1* or *REV -ERBα*), lysine-specific demethylase 1A (*KDM1A*), histone deacetylases (*HDACs*), ubiquitin ligases, F-box and leucine-rich protein repeat 3 (*FBXL3*), corepressor of nuclear receptor 2 (*NCOR2*), and nicotinamide phosphoribosyltransferase (*NAMPT*). These participate in various regulatory circuits designed to maintain the stability of the organism [105, 106].

There are reports describing different nutritional and environmental factors, including obesity, which can affect the DNA methylation pattern of clock genes that regulate the circadian rhythm in the hypothalamus and peripheral tissues. Insufficient sleep (short-term sleep or insomnia) has also been reported to be associated with loss of DNA methylation, which could be associated with alterations in pathways related to neuroplasticity, neurodegeneration, and cardiometabolic condition [107, 108].

Another neurological pathway of interest is the mesolimbic reward system. In this dopamine (DA) regulates pathological food intake. DA, which is synthesized from the amino acid tyrosine, exerts widespread effects both in neuronal tissues, as a neurotransmitter, and also in non-neuronal tissues as an autocrine or paracrine agent [109]. Dopamine reaches 80% of the total catecholamine levels in the mammalian brain. There are many subtypes of D1 and D2 receptors in the central nervous system; however, both types of dopamine receptors are implicated in neurobiological and behavioral disorders. There is evidence that food cues increase extracellular DA of the striatum. Therefore, dopamine plays a role in the non-hedonic motivational properties of food [110, 111]. Moreover, dopamine is suggested to encode the stimulatory properties of foods. The basis for this notion is that dopamine depletion or dopamine receptor blockade does not decrease pleasurable responses to palatable foods in animals or humans. People with obesity have been reported to have decreased levels of dopamine D2 receptors in the striatum, like observations in subjects with addictions. Dopamine deficiency can promote compensatory pathological eating to activate reward circuits [110, 112]. For instance, in 2013 it was suggested that epigenetic changes in adolescents contribute to long-lasting neurobiological consequences associated with early administration of ethanol (found in gasoline and adulterated alcoholic beverages) by causing brain region-specific changes in dopamine signaling [110].

4.3 Endocrine

The increased global prevalence of obesity is related with the use of industrial chemicals, and along the broad spectrum of these obesogenic elements, the endocrine disrupting compounds (EDCs) comprise up a sizable fraction [113]. EDCs were defined by the U.S. Environmental Protection Agency (EPA) as exogenous substances that disrupt the body's normal production, secretion, transport, metabolism, binding action, or elimination of blood-borne hormones that are necessary for homeostasis, reproduction, and developmental processes. Therefore, anormal exposure to EDCs, either natural or synthetic, impacts the hormonal and homeostatic systems which provide the organism the capacity to interact with its surroundings [114].

Several studies have considered dichlorodiphenyltrichloroethane (DDT) [115, 116] tributyltin (TBT) [117], diethylstilbestrol (DES) [118], perfluorooctanoic acid (PFOA) [119], and plastic derived like bisphenol-A (BPA), bis(2-ethylhexyl)phthalate (DEHP) and dibutyl phthalate (DBP) among others as EDCs [120]. Most of DES are molecular analogues of natural estrogens, having a high affinity for estrogenic and androgenic receptors (ERs). Additionally, it is currently known that the hypothalamus-pituitary gland-gonads (HPG), hypothalamus-pituitary gland-thyroid (HPT), and hypothalamus-pituitary gland-adrenal (HPA) are the principal EDC targets [121–123]. Hence, endocrine disruptors impair the endocrine and reproductive systems through a few mechanisms contributing to the global rise in diseases such as cancer, diabetes, neurological disorders and obesity [124–127].

Several works indicate that epigenetic alterations serve as a bridge between the harmful effects of EDCs and the onset of obesity in vulnerable individuals and during crucial developmental stages, starting from fetal life through infancy and puberty, even in pregnancy [28]. In this manner, and because of tissue accumulation and binding to hormone receptors, EDCs disrupt normal metabolic mechanisms altering adipose cells phenotype (increasing number and size) and adipocytokine (molecules primarily secreted by WA that function with paracrine and endocrine activity) production, therefore reducing basal metabolic rate and altering the control of satiety [122].

BPA is very used in industry for polycarbonate and plastics manufacturing, so it is one of the most studied EDC in both, *in vitro* and *in vivo* models [128]. By using a mouse BPA exposure model, it has been suggested that long-term BPA exposure increases lipid (including cholesterol) synthesis, improving hepatic lipid accumulation due to an hypomethylation over *Srebf1* and *Srebf2* [129]. These genes enhance its expression in early life comparable with aging animals, contributing to the early onset of metabolic disorders [129] like insulin resistance [130] linked to a reduction in adipokines (i.e., *ADIPOQ*, *FABP4*) [131] and adiponectin secretion and an elevated *resistin* expression [132]. Moreover, BPA can reduce global hepatic DNA methylation by a reduced activity of DNMTs as suggested by others [133]. Conversely, some other authors indicate that, although BPA can disturb lipid accumulation, it cannot affect adiponectin and leptin secretion [134], thus, the effects of BFA on endocrine systems require further investigations.

In addition, to elucidate how epigenetic mechanisms may explain the onset of obesity, to study the possible transgenerational actions of relevant EDCs has taken importance in last years. For example, the insecticide DDT has been used in rats to demonstrate that subsequent generations (F0: rats exposed to DDT & F3: great grand-offspring) can promote the obese phenotype (high body weight and abdominal adiposity) without a direct exposure, mainly in a critical gestating period [135]. As authors mentioned, this hereditary disease could be related with low CpG sites identified in F3 generation sperm compared with non-exposed male rats [135]. CpG demethylation characterize *leptin* promoter region and is highly expressed in differentiated adipocytes [136]. Obesity is followed by an increase in leptin secretion, modifying the action of leptin through the leptin receptor (LEPR) to reduce hunger by activating neurons that contain proopiomelanocortin (POMC) [137], which promoter has shown hypermethylation associated with 1) weight regain after dietary treatment [138] and 2) a high BMI [139].

4.4 Reproductive

Ovarian steroidal hormones control endometrial decidualization as a prerequisite for implantation during menstrual cycle trough epigenetic regulation. Bisphenol A (BPA) is a biologically compound active due to its property to bind directly with hormone receptors. BPA induce aberrant processes involved in reproduction and cells development at low concentrations. It is known that BPA has short lifespan within the organism, however, endometrial cells can retain BPA. Precisely, BPA-treated stromal cells reduced decidualization-related genes *PRL*, *IGFBP*-1 and *HOXA10*, related to a decrease of histone-3, lysine-4 trimethylation (H3K4me3) and an increase of histone-3, lysine-27 trimethylation (H3K27me3) in promoter regions. This effect was consisted with higher levels of *MLL1* and *EZH2*, histone methyltransferases responsible for the above histone modifications. These results suggested the epigenetic changes induced by BPA exposure impairing stromal cell decidualization, consequent embryo implantation failure and infertility [140]. Complementary studies demonstrated in human adipose tissue the enrichment of H3K4me3 histone mark in adipogenic, lipid metabolism and inflammatory promoter genes (*E2F1*, *LPL*, *SREBF2*, *SCD1*, *PPARG* and *IL6-IL9*) associated with Body Mass Index (BMI) and insulin resistance in morbid obese subjects with prediabetes compared to lean subjects [141]. Based on these findings, pollutants could induce metabolic deterioration by histone remodeling.

Bisphenol A (BPA) is widely used for polycarbonate plastics and resins production. Also, BPA is an agonist that binds with nuclear hormonal receptors causing endocrine disruption. Among BPA exposure consequences in female reproduction system are aberrant hypothalamic–pituitary hormonal production, oocyte quality reduction, defective uterine receptivity, polycystic ovary syndrome, premature puberty, and endometriosis development [142]. Di-(2-ethylhexyl) phthalate (DEHP) exposure affect male reproductive organ function and obesity triggering male secondary hypogonadism in High fat diet (HFD) mice due to oxidative stress induction [143].

A cohort study in women exposed to particulate matter with a diameter less than 2.5 μm (PM2.5, 13.4 μg/m3), black carbon (1.29 μg/m3), and nitrogen dioxide (17.98 μg/m3) during entire pregnancy demonstrated a positive correlation with cord blood histone H3 lysine 4 trimethylation (H3K4me3) levels. Precisely, this effect was significant after long exposure for black carbon and nitrogen dioxide, whereas cord blood histone H3 lysine 36 trimethylation (H3K36me3) levels inversely correlated in entire pregnancy, only for PM2.5 exposure [144]. In this regard, it is necessary to specify associated pathways to these histone modifications as potential markers for effective evaluation or even diagnosis in pregnant women in environmental risk zones.

Recent works have focused in elucidate ancestral environmental exposure on transgenerational epigenetic reprogramming of adipocytes. The methylation profile analysis in F3 generation of rats ancestrally exposed to DDT and atrazine demonstrated differential methylation regions (DMR) in 73% of genes (492/674), which corresponded between DDT obese male and atrazine lean male. The metabolic pathways related to these epigenetic changes was insulin signaling, adipogenesis, adipocyte browning, insulin resistance, and lipolysis. The adipocyte and metabolic genes in common between control and treatments were *Caln1, Ikzf1, Iqsec3, Kcnma1, Ksr2, Mycbp2, Myo16, Negr1, Nr1h5, Rbms3*, highly associated with obesity, type 2 diabetes, and metabolic syndrome predisposition. These results revealed that epigenetic alterations promoting by pollution contribute with obesity pathogenesis and metabolic diseases [145].

4.5 Immune

The fully functional immune system integrates organs, cells, pathways, and molecules in such an interconnected, sometimes circular, process that they often act synergistically to defend us from both internal and external aggressions [146]. When we refer to obesity and adverse environmental stimuli, such as pollution, from the immunological perspective we land in a crucial cellular process, inflammation. This process is promoted in the host through epigenetic alterations that mostly involve DNA methylation modifications [147]. Inflammation caused by adipose tissue in obese patients contributes to pathological conditions such as Diabetes Mellitus II (DM2) [148]. Recent studies have attempted to elucidate the specific role of B cells in obese patients after contrasting them with those in obese diabetic patients. Remarkably, the secretion of IL-6 and TNF-a increased in both groups, while a defect in the up-regulation of *IL-10* as well as higher concentrations of IgM and IgG was detected in obese diabetic subjects, in addition to poor response to stimulation with new antigens through vaccination against influenza [149]. This is extremely important as it shows the "snowball" effect in patients with these characteristics.

Adipose tissue is an immunologically active organ which contributes to systemic inflammation, especially the so-called white adipose tissue (WAT) in which subjects with obesity show phenotypic changes, due to inflammation, adipocyte dysfunctionality and infiltration of immune cells in vascular stromal fraction [150]. Furthermore, obesity involves chronic activation of the innate immune system and consequent local and systemic inflammation, activation of TP53 [151] and telomere shortening, a phenotype similar to aging [152]. In addition, an excess of fat mass is associated with several respiratory pathological conditions, such as asthma, obstructive sleep apnea, and chronic obstructive pulmonary disease [153]. It was recently found that obesity may be one of the missing pieces between pollution and severe clinical presentation in patients diagnosed with COVID-19 [154], representing another example of an additive action with environmental factors.

Air pollution exposure has been shown to increase the risk of obesity and metabolic dysfunctions in animal models and human studies [155]. An important finding is that higher exposure to near-roadway air pollution (NRAP), especially off-highway NRAP, was associated with higher concentrations of glycerol and metabolites related to oxidation of non-esterified fatty acids (NEFA). Besides this fact, plasma levels of NEFA and its oxidation by-products are associated with increased adiposity and insulin resistance. However, the association between air pollution exposure and serum NEFA concentration was not statistically significant, suggesting that the young population may still have adequate mitochondrial capacity to compensate environmental stressors [155]. In this line, NRAP exposure can stimulate pulmonary and systemic inflammation through activation of immunomodulatory receptors, such as Toll-like receptors (TLRs) [155].

5. Current and future epidrugs against environmental-pollution effects

Several studies support that aberrant epigenetic mechanisms are strongly related with human diseases, mainly by affecting protein expression and therefore the functionality of proteins that determine, in part, the epigenetic signature of diet-related diseases such as obesity [156]. Some of the most deregulated proteins are those encoded by leptin (*LEP*), melanocortin 4 receptor (*MC4R*), proopiomelanocortin (*POMC*) and insulin-like growth factor 2 (*IGF2*), as some causal genes for obesity due to epimutations. For instance, studies made in children and adults described a relationship between *POMC* intron hypermethylation (key element for food intake and energy balance regulation) and obesity. In addition, *IGF2* hypomethylation associates with higher BMI which also negatively correlates with specific lncRNAs levels like *lncRNA-p5549* and *lncRNA-p21015* [157].

It is necessary to remark that alterations in epigenetic landscapes are consequences of environmental factors like pharmacologicals, unhealthy habits, diet, and exposure to chemical stressors [157]. Certain exposure concentrations to chemical stressors or pollutants such as heavy metals, air pollution (PAHs, PM2.5, NO2) and EDCs during prenatal period are strongly related with high risk to overweight and obesity in childhood [158]. Metabolic disorders are also due to inflammatory responses characterized by immune cells infiltration promoted by pro-inflammatory cytokine release (i.e., TNF, IL-6, IL-1β, CCR2 and CCL2) [159]. The expression of inflammatory cytokines is also regulated by epigenomic regulators as in the case of CCL2 (CC chemokine ligand), which secretion is regulated by miRNAs in both positively (*miR-145*) and negatively (*miR-26a, miR - 92a, miR - 126, miR - 143, miR - 193a/b, miR - 652,* and miR*let-7a/d*) manners [159].

As mentioned, the diet is a pivotal factor for epigenetic reprogramming, thus, a healthy early-life dietary nutrition is crucial for the correct human developmental fate. Some bioactive dietary compounds largely studied in epigenetics are polyphenols and vitamins, also considered as phytochemicals, found in fruits, spices, teas and vegetables (**Figure 4**). They have been shown to have protective functions leading to

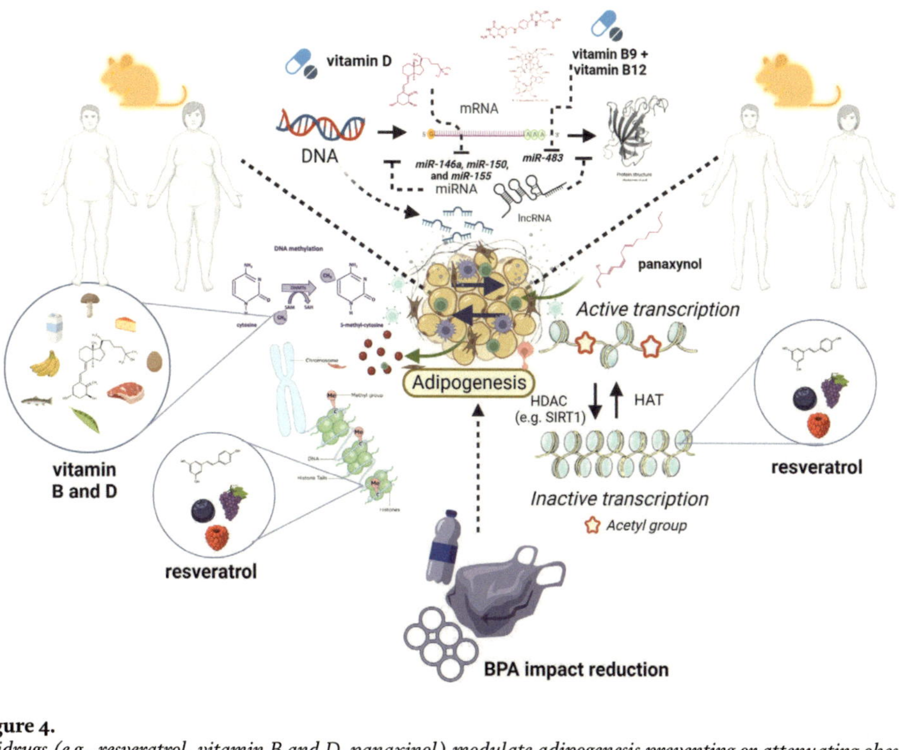

Figure 4.
Epidrugs (e.g., resveratrol, vitamin B and D, panaxinol) modulate adipogenesis preventing or attenuating obesity induced by pollutants. DNA: Desoxirribonucleic acid, SAM: S-adenosyl-methionine HDAC: Histone deacetylase, HAT: Histone acetylase, red dots: Adiponectin, vitamin D: Vitamin-D2 and -D3, 1α, 25-dihydroxyvitamin D3(1,25(OH)$_2$D3, bioactive form, calcitriol, vitamin B9: Folate and vitamin B12:Cobalamin.

healthy outcomes against to pollution-related diseases [160, 161]. Resveratrol (RSV) is a polyphenolic compound with anti-inflammatory, antioxidant and anticancer properties through DNA methylation and histone deacetylation modulation. In contrast, vitamin D has effects at DNA demethylation and histone acetylation levels promoting antitumoral processes [162]. Moreover, B vitamins are classified as methyl group donors, acting as precursors of SAM, therefore, when are included in diet they can counteract hypomethylation caused by BPA and diminish aberrant epigenetic inheritance [162].

Several bioactive dietary agents that potentially reverse epimutations and prevent or delay obesity, are of interest in the field of biomedicine because they can be propose as *epi-drugs* [163]. Due to their effects in modulating epigenetic mechanisms in human diseases, phytochemicals downregulate adipogenesis and upregulate fat oxidation preventing weight gain. For instance, resveratrol, which is found in plants, peanuts, berries, and grapes, can inhibit DNMTs activity and alter histone post-translational modifications by SIRT1 (HDAC Class III) activation [163]. In addition, it has been shown that resveratrol avoids adipocyte proliferation and preadipocyte differentiation through adipocyte-specific genes downregulation (*PPARγ*, *SREBP-1c*, *C/EBPa*, hormone-sensitive lipase, and lipoprotein lipase-*LPL*) in 3 T3-L1 adipocytes [164].

To ameliorate or to revert adipocyte differentiation could be an efficient therapy to inhibit the early onset of obesity-related diseases. In this sense, adiponectin, mainly secreted by white adipose tissue, has been proposed as a therapeutic target due to their crucial role in adipogenesis [165]. Adiponectin is secreted into the bloodstream as three oligomeric complexes, a trimer (67 kDa), a hexamer (140 kDa) and a high molecular weight (HMW, 300 kDa) formed by two trimers modified by hydroxylation and glycosylation, important for its metabolic-related downstream signaling functions [161]. A high circulating adiponectin expression protects against insulin resistance, alleviates lipotoxicity of lipid accumulation and it is associated with reduced age-related tissue inflammation [166], effects that are enhanced by peroxisome proliferator-activated receptor-γ PPARγ trough up-regulating *RUVBL2* (molecular chaperone for vesicle transport) expression as shown in 3 T3-L1 cells [167]. As mentioned, downregulation of *PPARγ* could diminish adipocyte differentiation, therefore it is crucial to identify effective agonists functioning as epidrugs that potentiate circulating adiponectin levels, such as panaxynol, isolated from *Saposhnikovia divaricata*, that can restore HMW adiponectin secretion affected by palmitic acid [168].

Nutriepigenomics is a promising field because gives important clinical information about certain ways to use phytochemichals and vitamins or to improve nano-engineering (vitamin formulations) as possible epidrugs to counteract the detrimental effects of environmental pollutants exposure (**Figure 4**) [169]. For this purpose, understanding the molecular mechanisms of vitamin-mediated epigenetic regulation by analyzing epidemiological, observational, *in vivo* and *in vitro* studies is crucial to identify effective bioactive compounds for human diseases treatments [170]. Vitamin D (VD) is a type of secosterol produced endogenously in the skin and obtained by diet, exists in two forms vitamin-D2 and -D3 and the bioactive form is 1α, 25-dihydroxyvitamin D3(1,25$(OH)_2$D3, calcitriol) involved in DNA methylation, in histone modifications and in miRNAs regulation [171]. Studies in mice have shown that a deficiency of maternal VD increases WA inflammatory responses in adults [172], thus, VD intake can limit inflammation and leukocyte infiltration by decreasing *miR-146a*, *miR-150*, and *miR-155* after a high fat (HF) diet supplemented with VD [173]. Additionally, combination of vitamins intake in defined doses could have greater benefits as epidrugs compared with single intake, since an adequate dietary of both vitamin B9 (folate) and vitamin B12 (cobalamin) may protect against metabolic imbalance [174]. As seen in female mice after fed them with both vitamins, expressed significant lower *miR-483* levels compared with mice fed with an altered dietary ratio, a high *miR-483* is linked to insulin resistance and DMT2 by impair lipid storage in adipose tissue [174]. Finally, both vitamins and phytochemicals (specially polyphenols) are now attracting the attention as the most promising bioactive compounds for a preventive anti-obesity and obese-related diseases therapy [160].

6. Conclusions and perspectives

"Globesity" has become a pandemic in the last years, adding a significant burden on the global health system. Although heritability of the disease is high and ethnicity-dependent, identified genetic variants associated with obesity represent a very small percentage of phenotypic variation. In recent years, epigenetic studies intend a deeper understanding of the molecular basis underlying the dramatic increase in global obesity rates. Existing evidence indicates that even environmental exposures induce alterations in the epigenome, leading to the transmission of obesity propensity across generations. In this chapter we have synthesized the effect of different pollution sources on epigenetic modulators. Therefore, as our understanding of epigenetics continues to advance, as do experimental approaches and sequencing techniques (already at the level of three-dimensional characterization of chromatin structure), biochemical and structural studies will become more imperative for unraveling novel catalytic-dependent and catalytic-independent functions of epigenetic regulators that together coordinate their actions on to induce diseases transcriptional landscapes.

Author details

Ivonne Ramirez-Diaz[1,2], Sagrario Lobato[1,3†], Andrea Perez-Gonzalez[1†], Alejandra Morales[1,4] and Karla Rubio[1†*]

1 International Laboratory EPIGEN, Consejo de Ciencia y Tecnología del Estado de Puebla (CONCYTEP), Instituto de Ciencias (ICUAP), Ecocampus Valsequillo, Benemérita Universidad Autónoma de Puebla (BUAP), Mexico

2 Universidad Popular Autónoma del Estado de Puebla (UPAEP), Mexico

3 Centro de Estudios Justo Sierra (CEJUS), Sinaloa, Mexico

4 Facultad de Medicina, Benemérita Universidad Autónoma de Puebla (BUAP), Puebla, Mexico

*Address all correspondence to: epigen.concytep@puebla.gob.mx

† Equal contribution.

References

[1] Chertow GM, Vart P, Jongs N, Langkilde AM, McMurray JJV, Correa-Rotter R, et al. Quetelet (body mass) index and effects of dapagliflozin in chronic kidney disease. Diabetes, Obesity & Metabolism. 2022;**24**(5):827-837

[2] Semlitsch T, Stigler FL, Jeitler K, Horvath K, Siebenhofer A. Management of overweight and obesity in primary care-a systematic overview of international evidence-based guidelines. Obesity Reviews. 2019;**20**(9):1218-1230

[3] Skolarczyk J, Pekar J, Łabądź D, Skórzyńska-Dziduszko K. Role of heavy metals in the development of obesity: A review of research. Journal of Elementology. 2018;**23**(4):1271-1280. DOI: 10.5601/jelem.2018.23.1.1545

[4] Smith JD, Fu E, Kobayashi MA. Prevention and Management of Childhood Obesity and its Psychological and health comorbidities. Annual Review of Clinical Psychology. 2020;**16**:351-378

[5] Littleton SH, Berkowitz RI, Grant SFA. Genetic determinants of childhood obesity. Molecular Diagnosis & Therapy. 2020;**24**(6):653-663

[6] Cuda SE, Censani M. Pediatric obesity algorithm: A practical approach to obesity diagnosis and management. Frontiers in Pediatrics. 2018;**6**:431

[7] Tyson N, Frank M. Childhood and adolescent obesity definitions as related to BMI, evaluation and management options. Best Practice & Research. Clinical Obstetrics & Gynaecology. 2018;**48**:158-164

[8] Kuhne SG, Stengel A. Alteration of peptidergic gut-brain signaling under conditions of obesity. Journal of Physiology and Pharmacology. 2019;**70**(5):651-665

[9] Pugliese G, Liccardi A, Graziadio C, Barrea L, Muscogiuri G, Colao A. Obesity and infectious diseases: Pathophysiology and epidemiology of a double pandemic condition. International Journal of Obesity. 2022;**46**(3):449-465

[10] Kivimaki M, Strandberg T, Pentti J, Nyberg ST, Frank P, Jokela M, et al. Body-mass index and risk of obesity-related complex multimorbidity: An observational multicohort study. The Lancet Diabetes and Endocrinology. 2022;**10**(4):253-263

[11] Bhaskaran K, Dos-Santos-Silva I, Leon DA, Douglas IJ, Smeeth L. Association of BMI with overall and cause-specific mortality: A population-based cohort study of 3.6 million adults in the UK. The Lancet Diabetes and Endocrinology. 2018;**6**(12):944-953

[12] Samodien E, Pheiffer C, Erasmus M, Mabasa L, Louw J, Johnson R. Diet-induced DNA methylation within the hypothalamic arcuate nucleus and dysregulated leptin and insulin signaling in the pathophysiology of obesity. Food Science & Nutrition. 2019;**7**(10):3131-3145

[13] Jaacks LM, Vandevijvere S, Pan A, McGowan CJ, Wallace C, Imamura F, et al. The obesity transition: Stages of the global epidemic. The Lancet Diabetes and Endocrinology. 2019;**7**(3):231-240

[14] Bray GA, Heisel WE, Afshin A, Jensen MD, Dietz WH, Long M, et al. The science of obesity management: An Endocrine Society scientific statement. Endocrine Reviews. 2018;**39**(2):79-132

[15] Di Ciaula A, Portincasa P. Diet and contaminants: Driving the rise to

obesity epidemics? Current Medicinal Chemistry. 2019;**26**(19):3471-3482

[16] Gadde KM, Martin CK, Berthoud HR, Heymsfield SB. Obesity: Pathophysiology and management. Journal of the American College of Cardiology. 2018;**71**(1):69-84

[17] Menting MD, Mintjens S, van de Beek C, Frick CJ, Ozanne SE, Limpens J, et al. Maternal obesity in pregnancy impacts offspring cardiometabolic health: Systematic review and meta-analysis of animal studies. Obesity Reviews. 2019;**20**(5):675-685

[18] Simeonov V. Environmental history of the twentieth century. An introductory didactic course. Chemistry-Didactics-Ecology-Metrology. 2019;**24**(1-2):7-21. DOI: 10.2478/cdem-2019-0001

[19] Khan MI, Chang Y-C. Environmental challenges and current practices in China—A thorough analysis. Sustainability. 2018;**10**(7):2547

[20] Manisalidis I, Stavropoulou E, Stavropoulos A, Bezirtzoglou E. Environmental and health impacts of air pollution: A review. Frontiers in Public Health. 2020;**8**:14

[21] Olmedo-Suarez MA, Ramirez-Diaz I, Perez-Gonzalez A, Molina-Herrera A, Coral-Garcia MA, Lobato S, et al. Epigenetic regulation in Exposome-induced tumorigenesis: Emerging roles of ncRNAs. Biomolecules. 2022;**12**(4):513

[22] WHO Global Air Quality Guidelines. Particulate Matter (PM 2.5 and PM 10), Ozone, Nitrogen Dioxide, Sulfur Dioxide and Carbon Monoxide. Geneva: World Health Organization; 2021 Licence: CC BY-NC-SA 3.0 IGO

[23] Available from: https://www.epa.gov/report-environment/indoor-air-quality

[24] An R, Ji M, Yan H, Guan C. Impact of ambient air pollution on obesity: A systematic review. International Journal of Obesity. 2018;**42**(6):1112-1126

[25] Mishra S, Dwivedi S, Singh R. A review on epigenetic effect of heavy metal carcinogens on human health. The Open Nutraceuticals Journal. 2010;**3**:188-193. DOI: 10.2174/1876396001003010188

[26] Pérez Castresana G, Castañeda Roldán E, García Suastegui WA, Morán Perales JL, Cruz Montalvo A, Handal SA. Evaluation of health risks due to heavy metals in a rural population exposed to Atoyac River pollution in Puebla, Mexico. Water. 2019;**11**(2):277

[27] Yang Z, Song Q, Li J, Zhang Y. Air pollution as a cause of obesity: Micro-level evidence from Chinese cities. International Journal of Environmental Research and Public Health. 2019;**16**(21):4296

[28] King SE, Skinner MK. Epigenetic transgenerational inheritance of obesity susceptibility. Trends in Endocrinology and Metabolism. 2020;**31**(7):478-494

[29] Waddington CH. Canalization of development and genetic assimilation of acquired characters. Nature. 1959;**183**(4676):1654-1655

[30] Halabian R, Valizadeh A, Ahmadi A, Saeedi P, Azimzadeh Jamalkandi S, Alivand MR. Laboratory methods to decipher epigenetic signatures: A comparative review. Cellular & Molecular Biology Letters. 2021;**26**(1):46

[31] Moore LD, Le T, Fan G. DNA methylation and its basic function. Neuropsychopharmacology. 2013;**38**(1):23-38

[32] Li E, Zhang Y. DNA methylation in mammals. Cold Spring Harbor Perspectives in Biology. 2014;**6**(5):a019133

[33] Manikkam M, Haque MM, Guerrero-Bosagna C, Nilsson EE, Skinner MK. Pesticide methoxychlor promotes the epigenetic transgenerational inheritance of adult-onset disease through the female germline. PLoS One. 2014;**9**(7):e102091

[34] King SE, McBirney M, Beck D, Sadler-Riggleman I, Nilsson E, Skinner MK. Sperm epimutation biomarkers of obesity and pathologies following DDT induced epigenetic transgenerational inheritance of disease. Environmental Epigenetics. 2019;**5**(2):dvz008

[35] Turner BM. Histone acetylation as an epigenetic determinant of long-term transcriptional competence. Cellular and Molecular Life Sciences. 1998;**54**(1):21-31

[36] Zhang T, Cooper S, Brockdorff N. The interplay of histone modifications - writers that read. EMBO Reports. 2015;**16**(11):1467-1481

[37] Arnaudo AM, Garcia BA. Proteomic characterization of novel histone post-translational modifications. Epigenetics & Chromatin. 2013;**6**(1):24

[38] Shi Y, Lan F, Matson C, Mulligan P, Whetstine JR, Cole PA, et al. Histone demethylation mediated by the nuclear amine oxidase homolog LSD1. Cell. 2004;**119**(7):941-953

[39] Ferrari A, Longo R, Peri C, Coppi L, Caruso D, Mai A, et al. Inhibition of class I HDACs imprints adipogenesis toward oxidative and brown-like phenotype. Biochimica et Biophysica Acta - Molecular and Cell Biology of Lipids. 2020;**1865**(4):158594

[40] Cricri D, Coppi L, Pedretti S, Mitro N, Caruso D, De Fabiani E, et al. Histone deacetylase 3 regulates adipocyte phenotype at early stages of differentiation. International Journal of Molecular Sciences. 2021;**22**(17):9300

[41] Patil VS, Zhou R, Rana TM. Gene regulation by non-coding RNAs. Critical Reviews in Biochemistry and Molecular Biology. 2014;**49**(1):16-32

[42] Taylor DH, Chu ET, Spektor R, Soloway PD. Long non-coding RNA regulation of reproduction and development. Molecular Reproduction and Development. 2015;**82**(12):932-956

[43] Akhade VS, Pal D, Kanduri C. Long noncoding RNA: Genome organization and mechanism of action. Advances in Experimental Medicine and Biology. 2017;**1008**:47-74

[44] Tiffon C. The impact of nutrition and environmental epigenetics on human health and disease. International Journal of Molecular Sciences. 2018;**19**(11):3425

[45] Srivastava S, Singh A, Sandeep K, Yadav D. Epigenetic regulation of gut microbial Dysbiosis. Indian Journal of Microbiology. 2021;**61**(2):125-129

[46] Pawel K, Maria MS. CpG Island Methylator phenotype-a Hope for the future or a road to nowhere? International Journal of Molecular Sciences. 2022;**23**(2):830

[47] Ortiz-Barahona V, Joshi RS, Esteller M. Use of DNA methylation profiling in translational oncology. Seminars in Cancer Biology. 2022;**83**:523-535

[48] Seale K, Horvath S, Teschendorff A, Eynon N, Voisin S. Making sense of the ageing methylome. Nature reviews. Genetics. 2022;**23**(10):585-605

[49] Hernando-Herraez I, Heyn H, Fernandez-Callejo M, Vidal E, Fernandez-Bellon H, Prado-Martinez J, et al. The interplay between DNA

methylation and sequence divergence in recent human evolution. Nucleic Acids Research. 2015;**43**(17):8204-8214

[50] Christiansen C, Tomlinson M, Eliot M, Nilsson E, Costeira R, Xia Y, et al. Adipose methylome integrative-omic analyses reveal genetic and dietary metabolic health drivers and insulin resistance classifiers. Genome Medicine. 2022;**14**(1):75

[51] Ray PD, Yosim A, Fry RC. Incorporating epigenetic data into the risk assessment process for the toxic metals arsenic, cadmium, chromium, lead, and mercury: Strategies and challenges. Frontiers in Genetics. 2014;**5**:201

[52] Chervona Y, Hall MN, Arita A, Wu F, Sun H, Tseng HC, et al. Associations between arsenic exposure and global posttranslational histone modifications among adults in Bangladesh. Cancer Epidemiology, Biomarkers & Prevention. 2012;**21**(12):2252-2260

[53] Hou L, Zhang X, Wang D, Baccarelli A. Environmental chemical exposures and human epigenetics. International Journal of Epidemiology. 2011;**41**(1):79-105

[54] Peng X, Yu S, Lin H, Wu F, Yang J, Zhou C, et al. Time-concentration-dependent profile of histone modifications on human hepatocytes treated by trichloroacetic acid. International Journal of Environmental Health Research. 2022;**32**(11):2376-2384

[55] Huguet A, Hatton A, Villot R, Quenault H, Blanchard Y, Fessard V. Modulation of chromatin remodelling induced by the freshwater cyanotoxin cylindrospermopsin in human intestinal caco-2 cells. PLoS One. 2014;**9**(6):e99121

[56] Burgio E, Lopomo A, Migliore L. Obesity and diabetes: From genetics to epigenetics. Molecular Biology Reports. 2015;**42**(4):799-818

[57] Kochumon S, Jacob T, Koshy M, Al-Rashed F, Sindhu S, Al-Ozairi E, et al. Palmitate potentiates lipopolysaccharide-induced IL-6 production via coordinated acetylation of H3K9/H3K18, p300, and RNA polymerase II. Journal of Immunology. 2022;**209**(4):731-741

[58] Iyer HS, Hart JE, Fiffer MR, Elliott EG, Yanosky JD, Kaufman JD, et al. Impacts of long-term ambient particulate matter and gaseous pollutants on circulating biomarkers of inflammation in male and female health professionals. Environmental Research. 2022;**214**(Pt 1):113810

[59] Humphrey KM, Pandey S, Martin J, Hagoel T, Grand'Maison A, Ohm JE. Establishing a role for environmental toxicant exposure induced epigenetic remodeling in malignant transformation. Seminars in Cancer Biology. 2019;**57**:86-94

[60] Lopez-Perez A, Norlin S, Steneberg P, Remeseiro S, Edlund H, Hornblad A. Pan-AMPK activator O304 prevents gene expression changes and remobilisation of histone marks in islets of diet-induced obese mice. Scientific Reports. 2021;**11**(1):24410

[61] Lu L, Hu J, Li G, An T. Low concentration Tetrabromobisphenol a (TBBPA) elevating overall metabolism by inducing activation of the Ras signaling pathway. Journal of Hazardous Materials. 2021;**416**:125797

[62] Ma Y, Li S, Ye S, Tang S, Hu D, Wei L, et al. Hexavalent chromium inhibits the formation of neutrophil extracellular traps and promotes the apoptosis of neutrophils via AMPK signaling pathway. Ecotoxicology and Environmental Safety. 2021;**223**:112614

[63] Gallego-Pauls M, Hernandez-Ferrer C, Bustamante M, Basagana X, Barrera-Gomez J, Lau CE, et al. Variability of multi-omics profiles in a

population-based child cohort. BMC Medicine. 2021;**19**(1):166

[64] Nurk S, Koren S, Rhie A, Rautiainen M, Bzikadze AV, Mikheenko A, et al. The complete sequence of a human genome. Science. 2022;**376**(6588):44-53

[65] Sarvari P, Sarvari P, Ramirez-Diaz I, Mahjoubi F, Rubio K. Advances of epigenetic biomarkers and Epigenome editing for early diagnosis in breast cancer. International Journal of Molecular Sciences. 2022;**23**(17):9521

[66] Rubio K, Singh I, Dobersch S, Sarvari P, Gunther S, Cordero J, et al. Inactivation of nuclear histone deacetylases by EP300 disrupts the MiCEE complex in idiopathic pulmonary fibrosis. Nature Communications. 2019;**10**(1):2229

[67] Singh I, Contreras A, Cordero J, Rubio K, Dobersch S, Gunther S, et al. MiCEE is a ncRNA-protein complex that mediates epigenetic silencing and nucleolar organization. Nature Genetics. 2018;**50**(7):990-1001

[68] Castanotto D, Lingeman R, Riggs AD, Rossi JJ. CRM1 mediates nuclear-cytoplasmic shuttling of mature microRNAs. Proceedings of the National Academy of Sciences of the United States of America. 2009;**106**(51):21655-21659

[69] Reyes-Gutierrez P, Ritland Politz JC, Pederson T. A mRNA and cognate microRNAs localize in the nucleolus. Nucleus. 2014;**5**(6):636-642

[70] Rubio K, Castillo-Negrete R, Barreto G. Non-coding RNAs and nuclear architecture during epithelial-mesenchymal transition in lung cancer and idiopathic pulmonary fibrosis. Cellular Signalling. 2020;**70**:109593

[71] Landrier JF, Derghal A, Mounien L. MicroRNAs in obesity and related metabolic disorders. Cell. 2019;**8**(8):859

[72] Kunej T, Jevsinek Skok D, Zorc M, Ogrinc A, Michal JJ, Kovac M, et al. Obesity gene atlas in mammals. Journal of Genomics. 2013;**1**:45-55

[73] Formichi C, Nigi L, Grieco GE, Maccora C, Fignani D, Brusco N, et al. Non-coding RNAs: Novel players in insulin resistance and related diseases. International Journal of Molecular Sciences. 2021;**22**(14):7716

[74] Grun F, Blumberg B. Environmental obesogens: Organotins and endocrine disruption via nuclear receptor signaling. Endocrinology. 2006;**147**(6 Suppl):S50-S55

[75] Mahmoud AM. An overview of epigenetics in obesity: The role of lifestyle and therapeutic interventions. International Journal of Molecular Sciences. 2022;**23**(3):1341

[76] Efsa Panel on Food Contact Materials E, Processing A, Silano V, Barat Baviera JM, Bolognesi C, Chesson A, et al. Update of the risk assessment of di-butylphthalate (DBP), butyl-benzyl-phthalate (BBP), bis(2-ethylhexyl) phthalate (DEHP), di-isononylphthalate (DINP) and di-isodecylphthalate (DIDP) for use in food contact materials. EFSA Journal. 2019;**17**(12):e05838

[77] Meruvu S, Zhang J, Choudhury M. Butyl benzyl phthalate promotes Adipogenesis in 3T3-L1 cells via the miRNA-34a-5p signaling pathway in the absence of exogenous Adipogenic stimuli. Chemical Research in Toxicology. 2021;**34**(11):2251-2260

[78] McIlwraith EK, Lieu CV, Belsham DD. Bisphenol a induces miR-708-5p through an ER stress-mediated mechanism altering neuronatin and neuropeptide Y expression in hypothalamic neuronal models. Molecular and Cellular Endocrinology. 2022;**539**:111480

[79] Rahmani S, Vakhshiteh F, Hodjat M, Sahranavardfard P, Hassani S, Ghafour Broujerdi E, et al. Gene-environmental interplay in Bisphenol a subchronic animal exposure: New insights into the epigenetic regulation of pancreatic islets. Chemical Research in Toxicology. 2020;**33**(9):2338-2350

[80] Verbanck M, Canouil M, Leloire A, Dhennin V, Coumoul X, Yengo L, et al. Low-dose exposure to bisphenols a, F and S of human primary adipocyte impacts coding and non-coding RNA profiles. PLoS One. 2017;**12**(6):e0179583

[81] Pergoli L, Cantone L, Favero C, Angelici L, Iodice S, Pinatel E, et al. Extracellular vesicle-packaged miRNA release after short-term exposure to particulate matter is associated with increased coagulation. Particle and Fibre Toxicology. 2017;**14**(1):32

[82] Park YJ, Han SM, Huh JY, Kim JB. Emerging roles of epigenetic regulation in obesity and metabolic disease. The Journal of Biological Chemistry. 2021;**297**(5):101296

[83] Su C, Gao L, May CL, Pippin JA, Boehm K, Lee M, et al. 3D chromatin maps of the human pancreas reveal lineage-specific regulatory architecture of T2D risk. Cell Metabolism. 2022;**34**(9):1394-409 e4

[84] Chen L, Li J, Yuan R, Wang Y, Zhang J, Lin Y, et al. Dynamic 3D genome reorganization during development and metabolic stress of the porcine liver. Cell Discovery. 2022;**8**(1):56

[85] Qin Y, Grimm SA, Roberts JD, Chrysovergis K, Wade PA. Alterations in promoter interaction landscape and transcriptional network underlying metabolic adaptation to diet. Nature Communications. 2020;**11**(1):962

[86] Wozniak E, Reszka E, Jablonska E, Michalowicz J, Huras B, Bukowska B. Glyphosate and AMPA induce alterations in expression of genes involved in chromatin architecture in human peripheral blood mononuclear cells (In vitro). International Journal of Molecular Sciences. 2021;**22**(6):2966

[87] Prinz N, Korez Š. Understanding how microplastics affect marine biota on the cellular level is important for assessing ecosystem function: A review. In: Jungblut S, Liebich V, Bode-Dalby M, editors. YOUMARES 9 - the Oceans: Our Research, our Future. Cham: Springer; 2020. DOI: 10.1007/978-3-030-20389-4_6

[88] World Health Organization. Regional Office for Europe. Health Risks of Persistent Organic Pollutants from Long-Range Transboundary Air Pollution. Copenhagen: WHO Regional Office for Europe; 2003

[89] Available from: https://www.who.int/news-room/fact-sheets/detail/drinking-water

[90] Charette T, Kaminski G, Rosabal M, Amyot M. Effects of speciation, cooking and changes in bioaccessibility on methylmercury exposure assessment for contrasting diets of fish and marine mammals. International Journal of Environmental Research and Public Health. 2021;**18**(5):2565

[91] Berg G, Rybakova D, Fischer D, Cernava T, Verges MC, Charles T, et al. Microbiome definition re-visited: Old concepts and new challenges. Microbiome. 2020;**8**(1):103

[92] Dhar D, Mohanty A. Gut microbiota and Covid-19- possible link and implications. Virus Research. 2020;**285**:198018

[93] Rinninella E, Raoul P, Cintoni M, Franceschi F, Miggiano GAD,

Gasbarrini A, et al. What is the healthy gut microbiota composition? A changing ecosystem across age, environment, diet, and diseases. Microorganisms. 2019;7(1):14

[94] Rowland I, Gibson G, Heinken A, Scott K, Swann J, Thiele I, et al. Gut microbiota functions: Metabolism of nutrients and other food components. European Journal of Nutrition. 2018;**57**(1):1-24

[95] Akhtar M, Chen Y, Ma Z, Zhang X, Shi D, Khan JA, et al. Gut microbiota-derived short chain fatty acids are potential mediators in gut inflammation. Animal Nutrition. 2022;**8**:350-360

[96] Pulliero A, Traversi D, Franchitti E, Barchitta M, Izzotti A, Agodi A. The interaction among microbiota, epigenetic regulation, and air pollutants in disease prevention. Journal of Personalized Medicine. 2021;**12**(1):14

[97] Wang W, Zhou J, Chen M, Huang X, Xie X, Li W, et al. Exposure to concentrated ambient PM2.5 alters the composition of gut microbiota in a murine model. Particle and Fibre Toxicology. 2018;**15**(1):17

[98] Sharma M, Li Y, Stoll ML, Tollefsbol TO. The epigenetic connection between the gut microbiome in obesity and diabetes. Frontiers in Genetics. 2019;**10**:1329

[99] Plano SA, Casiraghi LP, Garcia Moro P, Paladino N, Golombek DA, Chiesa JJ. Circadian and metabolic effects of light: Implications in weight homeostasis and health. Frontiers in Neurology. 2017;**8**:558

[100] Walker WH 2nd, Walton JC, DeVries AC, Nelson RJ. Circadian rhythm disruption and mental health. Translational Psychiatry. 2020;**10**(1):28

[101] Karatsoreos IN. Circadian regulation of the brain and behavior: A neuroendocrine perspective. Current Topics in Behavioral Neurosciences. 2019;**43**:323-351

[102] Kumar P, Ashawat SM, Pandit V, Sharma KD. Artificial light pollution at night: A risk for Normal circadian rhythm and physiological functions in humans, current. Environmental Engineering. 2019;**6**(2):111-125. DOI: 10.2174/2212717806666190619120211

[103] Qin T, Zhang F, Zhou H, Ren H, Du Y, Liang S, et al. High-level PM2.5/PM10 exposure is associated with alterations in the human pharyngeal microbiota composition. Frontiers in Microbiology. 2019;**10**:54

[104] Fan Y, Pedersen O. Gut microbiota in human metabolic health and disease. Nature reviews. Microbiology. 2021;**19**(1):55-71

[105] Hudec M, Dankova P, Solc R, Bettazova N, Cerna M. Epigenetic regulation of circadian rhythm and its possible role in diabetes mellitus. International Journal of Molecular Sciences. 2020;**21**(8):3005

[106] Zou X, Kim DW, Gotoh T, Liu J, Kim JK, Finkielstein CV. A systems biology approach identifies hidden regulatory connections between the circadian and cell-cycle checkpoints. Frontiers in Physiology. 2020;**11**:327

[107] Samblas M, Milagro FI, Martínez A. DNA methylation markers in obesity, metabolic syndrome, and weight loss. Epigenetics. 2019;**14**(5):421-444

[108] Fagiani F, Di Marino D, Romagnoli A, Travelli C, Voltan D, Di Cesare ML, et al. Molecular regulations of circadian rhythm and implications for physiology and diseases. Signal

Transduction and Targeted Therapy. 2022;**7**(1):41

[109] Medina-Pérez G, Fernández-Luqueño F, Vazquez-Nuñez E, López-Valdez F, Prieto-Mendez J, Madariaga-Navarrete A, et al. Remediating polluted soils using nanotechnologies: Environmental benefits and risks. Polish Journal of Environmental Studies. 2019;**28**(3):1013-1030. DOI: 10.15244/pjoes/87099

[110] Osinga TE, Links TP, Dullaart RPF, Pacak K, van der Horst-Schrivers ANA, Kerstens MN, et al. Emerging role of dopamine in neovascularization of pheochromocytoma and paraganglioma. The FASEB Journal. 2017;**31**(6):2226-2240

[111] Melis MR, Sanna F, Argiolas A. Dopamine, erectile function and male sexual behavior from the past to the present: A review. Brain Sciences. 2022;**12**(7):826

[112] Rampin O, Saint Albin Deliot A, Ouali C, Burguet J, Gry E, Champeil Potokar G, et al. Dopamine modulates the processing of food odour in the ventral striatum. Biomedicine. 2022;**10**(5):1126

[113] Wang G, Walker SO, Hong X, Bartell TR, Wang X. Epigenetics and early life origins of chronic noncommunicable diseases. The Journal of Adolescent Health. 2013;**52**(2 Suppl 2):S14-S21

[114] Diamanti-Kandarakis E, Bourguignon JP, Giudice LC, Hauser R, Prins GS, Soto AM, et al. Endocrine-disrupting chemicals: An Endocrine Society scientific statement. Endocrine Reviews. 2009;**30**(4):293-342

[115] La Merrill MA, Krigbaum NY, Cirillo PM, Cohn BA. Association between maternal exposure to the pesticide dichlorodiphenyltrichloroethane (DDT) and risk of obesity in middle age. International Journal of Obesity. 2020;**44**(8):1723-1732

[116] Tawar N, Banerjee BD, Mishra BK, Sharma T, Tyagi S, Madhu SV, et al. Adipose tissue levels of DDT as risk factor for obesity and type 2 diabetes mellitus. Indian Journal of Endocrinology and Metabolism. 2021;**25**(2):160-165

[117] Li X, Ycaza J, Blumberg B. The environmental obesogen tributyltin chloride acts via peroxisome proliferator activated receptor gamma to induce adipogenesis in murine 3T3-L1 preadipocytes. The Journal of Steroid Biochemistry and Molecular Biology. 2011;**127**(1-2):9-15

[118] Owuor TO, Reid M, Reschke L, Hagemann I, Greco S, Modi Z, et al. Maternal obesogenic diet induces endometrial hyperplasia, an early hallmark of endometrial cancer, in a diethylstilbestrol mouse model. PLoS One. 2018;**13**(5):e0186390

[119] Shao W, Xu J, Xu C, Weng Z, Liu Q, Zhang X, et al. Early-life perfluorooctanoic acid exposure induces obesity in male offspring and the intervention role of chlorogenic acid. Environmental Pollution. 2021;**272**:115974

[120] Manikkam M, Tracey R, Guerrero-Bosagna C, Skinner MK. Plastics derived endocrine disruptors (BPA, DEHP and DBP) induce epigenetic transgenerational inheritance of obesity, reproductive disease and sperm epimutations. PLoS One. 2013;**8**(1):e55387

[121] Yang Q, Yang X, Liu J, Ren W, Chen Y, Shen S. Effects of BPF on steroid hormone homeostasis and gene expression in the hypothalamic-pituitary-gonadal axis of zebrafish. Environmental Science and Pollution Research International. 2017;**24**(26):21311-21322

[122] Wu H, Zhang W, Zhang Y, Kang Z, Miao X, Na X. Novel insights into di(2ethylhexyl)phthalate activation: Implications for the hypothalamuspituitarythyroid axis. Molecular Medicine Reports. 2021;**23**(4):290

[123] Lauretta R, Sansone A, Sansone M, Romanelli F, Appetecchia M. Endocrine disrupting chemicals: Effects on endocrine glands. Frontiers in Endocrinology (Lausanne). 2019;**10**:178

[124] Zhuang S, Zhang J, Wen Y, Zhang C, Liu W. Distinct mechanisms of endocrine disruption of DDT-related pesticides toward estrogen receptor alpha and estrogen-related receptor gamma. Environmental Toxicology and Chemistry. 2012;**31**(11):2597-2605

[125] Adam AHB, de Haan LHJ, Estruch IM, Hooiveld G, Louisse J, Rietjens I. Estrogen receptor alpha (ERalpha)-mediated coregulator binding and gene expression discriminates the toxic ERalpha agonist diethylstilbestrol (DES) from the endogenous ERalpha agonist 17beta-estradiol (E2). Cell Biology and Toxicology. 2020;**36**(5):417-435

[126] Sakai G, Hirao-Suzuki M, Koga T, Kobayashi T, Kamishikiryo J, Tanaka M, et al. Perfluorooctanoic acid (PFOA) as a stimulator of estrogen receptor-negative breast cancer MDA-MB-231 cell aggressiveness: Evidence for involvement of fatty acid 2-hydroxylase (FA2H) in the stimulated cell migration. The Journal of Toxicological Sciences. 2022;**47**(4):159-168

[127] Li Y, Luh CJ, Burns KA, Arao Y, Jiang Z, Teng CT, et al. Endocrine-disrupting chemicals (EDCs): In vitro mechanism of estrogenic activation and differential effects on ER target genes. Environmental Health Perspectives. 2013;**121**(4):459-466

[128] Desai M, Ferrini MG, Jellyman JK, Han G, Ross MG. In vivo and in vitro bisphenol a exposure effects on adiposity. Journal of Developmental Origins of Health and Disease. 2018;**9**(6):678-687

[129] Ke ZH, Pan JX, Jin LY, Xu HY, Yu TT, Ullah K, et al. Bisphenol a exposure May induce hepatic lipid accumulation via reprogramming the DNA methylation patterns of genes involved in lipid metabolism. Scientific Reports. 2016;**6**:31331

[130] Carlsson A, Sorensen K, Andersson AM, Frederiksen H, Juul A. Bisphenol a, phthalate metabolites and glucose homeostasis in healthy normal-weight children. Endocrine Connections. 2018;**7**(1):232-238

[131] Ahmed F, Sarsenbayeva A, Katsogiannos P, Aguer C, Pereira MJ. The effects of bisphenol a and bisphenol S on adipokine expression and glucose metabolism in human adipose tissue. Toxicology. 2020;**445**:152600

[132] Nettore IC, Franchini F, Palatucci G, Macchia PE, Ungaro P. Epigenetic mechanisms of endocrine-disrupting Chemicals in Obesity. Biomedicine. 2021;**9**(11):1716

[133] Ma Y, Xia W, Wang DQ, Wan YJ, Xu B, Chen X, et al. Hepatic DNA methylation modifications in early development of rats resulting from perinatal BPA exposure contribute to insulin resistance in adulthood. Diabetologia. 2013;**56**(9):2059-2067

[134] Szkudelska K, Okulicz M, Szkudelski T. Bisphenol a disturbs metabolism of primary rat adipocytes without affecting adipokine secretion. Environmental Science and Pollution Research International. 2021;**28**(18):23301-23309

[135] Skinner MK, Manikkam M, Tracey R, Guerrero-Bosagna C, Haque M, Nilsson EE. Ancestral dichlorodiphenyltrichloroethane (DDT) exposure promotes epigenetic transgenerational inheritance of obesity. BMC Medicine. 2013;**11**:228

[136] Melzner I, Scott V, Dorsch K, Fischer P, Wabitsch M, Bruderlein S, et al. Leptin gene expression in human preadipocytes is switched on by maturation-induced demethylation of distinct CpGs in its proximal promoter. The Journal of Biological Chemistry. 2002;**277**(47):45420-45427

[137] Wroblewski A, Strycharz J, Swiderska E, Drewniak K, Drzewoski J, Szemraj J, et al. Molecular insight into the interaction between epigenetics and leptin in metabolic disorders. Nutrients. 2019;**11**(8):1872

[138] Crujeiras AB, Campion J, Diaz-Lagares A, Milagro FI, Goyenechea E, Abete I, et al. Association of weight regain with specific methylation levels in the NPY and POMC promoters in leukocytes of obese men: A translational study. Regulatory Peptides. 2013;**186**:1-6

[139] Dunstan J, Bressler JP, Moran TH, Pollak JS, Hirsch AG, Bailey-Davis L, et al. Associations of LEP, CRH, ICAM-1, and LINE-1 methylation, measured in saliva, with waist circumference, body mass index, and percent body fat in mid-childhood. Clinical Epigenetics. 2017;**9**:29

[140] Xiong Y, Wen X, Liu H, Zhang M, Zhang Y. Bisphenol a affects endometrial stromal cells decidualization, involvement of epigenetic regulation. The Journal of Steroid Biochemistry and Molecular Biology. 2020;**200**:105640

[141] Castellano-Castillo D, Denechaud PD, Fajas L, Moreno-Indias I, Oliva-Olivera W, Tinahones F, et al. Human adipose tissue H3K4me3 histone mark in adipogenic, lipid metabolism and inflammatory genes is positively associated with BMI and HOMA-IR. PLoS One. 2019;**14**(4):e0215083

[142] Caserta D, Di Segni N, Mallozzi M, Giovanale V, Mantovani A, Marci R, et al. Bisphenol a and the female reproductive tract: An overview of recent laboratory evidence and epidemiological studies. Reproductive Biology and Endocrinology. 2014;**12**:37

[143] Zhao J, Ren S, Liu C, Huo L, Liu Z, Zhai L. Di-(2-Ethylhexyl) phthalate increases obesity-induced damage to the male reproductive system in mice. Oxidative Medicine and Cellular Longevity. 2018;**2018**:1861984

[144] Vrijens K, Trippas AJ, Lefebvre W, Vanpoucke C, Penders J, Janssen BG, et al. Association of Prenatal Exposure to ambient air pollution with circulating histone levels in maternal cord blood. JAMA Network Open. 2020;**3**(5):e205156

[145] King SE, Nilsson E, Beck D, Skinner MK. Adipocyte epigenetic alterations and potential therapeutic targets in transgenerationally inherited lean and obese phenotypes following ancestral exposures. Adipocytes. 2019;**8**(1):362-378

[146] Punt J, Stranford SA, Jones PP, Owen JA, editors. Visión general del sistema inmune. In: KUBY. Inmunología. 8e ed. McGraw Hill Interamericana Editores, S.A. de C.V. 2020

[147] Ramos-Lopez O, Milagro FI, Riezu-Boj JI, Martinez JA. Epigenetic signatures underlying inflammation: An interplay of nutrition, physical activity, metabolic diseases, and environmental factors for personalized nutrition. Inflammation Research. 2021;**70**(1):29-49

[148] Russo S, Kwiatkowski M, Govorukhina N, Bischoff R, Melgert BN.

Meta-inflammation and metabolic reprogramming of macrophages in diabetes and obesity: The importance of metabolites. Frontiers in Immunology. 2021;**12**:746151

[149] Zhai X, Qian G, Wang Y, Chen X, Lu J, Zhang Y, et al. Elevated B cell activation is associated with type 2 diabetes development in obese subjects. Cellular Physiology and Biochemistry. 2016;**38**(3):1257-1266

[150] Kawai T, Autieri MV, Scalia R. Adipose tissue inflammation and metabolic dysfunction in obesity. American Journal of Physiology. Cell Physiology. 2021;**320**(3):C375-CC91

[151] Zwezdaryk K, Sullivan D, Saifudeen Z. The p53/adipose-tissue/cancer nexus. Frontiers in Endocrinology (Lausanne). 2018;**9**:457

[152] Frasca D, Diaz A, Romero M, Blomberg BB. Ageing and obesity similarly impair antibody responses. Clinical and Experimental Immunology. 2017;**187**(1):64-70

[153] Peters U, Suratt BT, Bates JHT, Dixon AE. Beyond BMI: Obesity and lung disease. Chest. 2018;**153**(3):702-709

[154] Lubrano C, Risi R, Masi D, Gnessi L, Colao A. Is obesity the missing link between COVID-19 severity and air pollution? Environmental Pollution. 2020;**266**(Pt 3):115327

[155] Chen Z, Newgard CB, Kim JS O II, Alderete TL, Thomas DC, et al. Near-roadway air pollution exposure and altered fatty acid oxidation among adolescents and young adults - the interplay with obesity. Environment International 2019;130:104935.

[156] Bouyahya A, Mechchate H, Oumeslakht L, Zeouk I, Aboulaghras S, Balahbib A, et al. The role of epigenetic modifications in human cancers and the use of natural compounds as Epidrugs: Mechanistic pathways and Pharmacodynamic actions. Biomolecules. 2022;**12**(3):367

[157] Obri A, Serra D, Herrero L, Mera P. The role of epigenetics in the development of obesity. Biochemical Pharmacology. 2020;**177**:113973

[158] Simkova S, Veleminsky M, Sram RJ. The impact of air pollution to obesity. Neuro Endocrinology Letters. 2020;**41**(3):146-153

[159] Jung BC, Kang S. Epigenetic regulation of inflammatory factors in adipose tissue. Biochimica et Biophysica Acta - Molecular and Cell Biology of Lipids. 2021;**1866**(11):159019

[160] Boughanem H, Kompella P, Tinahones FJ, Macias-Gonzalez M. An overview of vitamins as epidrugs for colorectal cancer prevention. Nutrition Reviews. 2022;**81**(4):455-479

[161] Ratre P, Chauhan P, Bhargava A, Tiwari R, Thareja S, Srivastava RK, et al. Nano-engineered vitamins as a potential epigenetic modifier against environmental air pollutants. Reviews on Environmental Health. 2022. DOI: 10.1515/reveh-2022-0027

[162] Takagi M, Kimura K, Nakashima KI, Inoue M. Ameliorative effect of panaxynol on the reduction in high-molecular-weight adiponectin secretion from 3T3-L1 adipocytes treated with palmitic acids. European Journal of Pharmacology. 2018;**820**:138-145

[163] Zhu D, Xu L, Wei X, Xia B, Gong Y, Li Q, et al. PPARgamma enhanced adiponectin polymerization and trafficking by promoting RUVBL2 expression during adipogenic differentiation. Gene. 2021;**764**:145100

[164] Li N, Zhao S, Zhang Z, Zhu Y, Gliniak CM, Vishvanath L, et al. Adiponectin preserves metabolic fitness during aging. eLife. 2021;**10**:e65108

[165] Achari AE, Jain SK. Adiponectin, a therapeutic target for obesity, diabetes, and endothelial dysfunction. International Journal of Molecular Sciences. 2017;**18**(6):1321

[166] Rayalam S, Yang JY, Ambati S, Della-Fera MA, Baile CA. Resveratrol induces apoptosis and inhibits adipogenesis in 3T3-L1 adipocytes. Phytotherapy Research. 2008;**22**(10):1367-1371

[167] Dincer Y, Yuksel S. Antiobesity effects of phytochemicals from an epigenetic perspective. Nutrition. 2021;**84**:111119

[168] Ramirez-Alarcon K, Victoriano M, Mardones L, Villagran M, Al-Harrasi A, Al-Rawahi A, et al. Phytochemicals as potential Epidrugs in type 2 diabetes mellitus. Frontiers in Endocrinology (Lausanne). 2021;**12**:656978

[169] Li S, Chen M, Li Y, Tollefsbol TO. Prenatal epigenetics diets play protective roles against environmental pollution. Clinical Epigenetics. 2019;**11**(1):82

[170] Saad B, Ghareeb B, Kmail A. Metabolic and epigenetics action mechanisms of Antiobesity medicinal plants and phytochemicals. Evidence-based Complementary and Alternative Medicine. 2021;**2021**:9995903

[171] Mahajan A, Sapehia D, Thakur S, Mohanraj PS, Bagga R, Kaur J. Effect of imbalance in folate and vitamin B12 in maternal/parental diet on global methylation and regulatory miRNAs. Scientific Reports. 2019;**9**(1):17602

[172] Karkeni E, Bonnet L, Marcotorchino J, Tourniaire F, Astier J, Ye J, et al. Vitamin D limits inflammation-linked microRNA expression in adipocytes in vitro and in vivo: A new mechanism for the regulation of inflammation by vitamin D. Epigenetics. 2018;**13**(2):156-162

[173] Haroun N, Bennour I, Seipelt E, Astier J, Couturier C, Mounien L, et al. Maternal vitamin D deficiency in mice increases white adipose tissue inflammation in offspring. Cell. 2022;**11**(13):2024

[174] Nur SM, Rath S, Ahmad V, Ahmad A, Ateeq B, Khan MI. Nutritive vitamins as epidrugs. Critical Reviews in Food Science and Nutrition. 2021;**61**(1):1-13

Chapter 5

Epigenetics in Reproductive Aging: Involvement of Oxidative Stress

Olalekan Bukunmi Ogunro and Oladimeji Taiwo Babatunde

Abstract

Epigenetic alterations are one of the distinctive characteristics of aging. Epigenetics changes throughout the reproductive life of humans. The major epigenetic parameters viz. DNA methylation, histone modification, and chromatin modeling are altered in the oocyte and sperm due to aging. Also, aging is accompanied by oxidative stress resulting in oocyte and sperm DNA damage. Oxidative stress occurs when the body's antioxidant defense mechanism is overwhelmed by free radicals or pro-oxidant molecules such as nitrogen and reactive oxygen species, which are generated during normal cellular metabolism. This phenomenon is accompanied by a decline in the cell repair machinery, resulting in a wide range of DNA damage and distortion in cellular epigenetics. Still more, free radicals can directly or indirectly interfere with some epigenetic processes of the cell. For example, free radicals can impart the genome methylation profile by forming oxidized DNA lesions. Given the deleterious impact of oxidative stress on aging and cellular epigenetic profile, the ingestion of external antioxidants is encouraged to circumvent its side effects. This chapter provides insight into the interconnection between epigenetic alterations (histone modification, chromatin remodeling, DNA methylation and miRNA), reproductive aging, and oxidative stress.

Keywords: epigenetics, aging, reproduction, oxidative stress, DNA methylation, histone modification

1. Introduction

God blessed them, and God said to them, 'Be fruitful and multiply, and fill the earth' (Gen. 1:28a). Reproduction, a vital characteristic of life is a biological process by which new organisms (offspring) are made from their parents. Simply put, organisms replicate themselves via reproduction either sexually or asexually. For sexual reproduction, a new individual is made by the combination (union) of two gametes or sex cells - sperm and eggs (or ova) while in asexual reproduction, organisms make genetically identical or similar copies of themselves without contributing any genetic resources from another organism. In humans, reproductive functions diminish with age. It stops abruptly in females at menopause while in males it gradually declines. Aging is therefore an unavoidable and unappeasable natural process markedly disproportional gender-based impressions on human fertility. Epigenetic alterations are one of the distinctive characteristics of aging, epigenetics changes throughout the reproductive life in human. In human, oxidative stress (described when there is an imbalance between the antioxidant defense system and pro-oxidant molecules - nitrogen and reactive oxygen species) play a major role in both male and female reproduction. Oxidative stress has adverse effects on both female and male gametes and the developmental capacity of embryos [1]. For instance, sperm quality and functions in males (characterized by immature and morphologically abnormal spermatozoa as well as infected white blood cells in the seminal ejaculates) are adversely affected in conditions of oxidative stress while in females, oxidative stress is involved in a number of age-related reproductive diseases such as endometriosis, tubal and peritoneal factor infertility, polycystic ovary syndrome (PCOS), ovarian cancers, and unexplained infertility. Furthermore, complications from pregnancy such as recurrent pregnancy loss, spontaneous abortion/still birth, hydatidiform mole, and preeclampsia have been linked to oxidative stress in females [2, 3]. All these alterations (aging) in the reproduction characterized by a progressive loss in both physiological and cellular functions are possible because of the enforced aging caused by oxidation stress that may be induced by various social habits and environmental factors [2]. The role of epigenetics has been suggested as one of the molecular mechanisms linking stress and aging by manipulating genomic function and phenotypic composition, such as aging-related consequences. Moreover, epigenetic modifications (covalent modifications of DNA and key histones that modulate activity of gene without altering the sequence of DNA) and chromatins aggregation play a vital role in reproductive aging [4]. For instance, in male, epigenetic mechanisms involved in the regulation of oxidative stress in the male reproductive system are crucial in spermatogenesis to keep testicular homeostasis through the modulation of molecular pathways. Also in females, alterations in epigenetics and its associated enzymes such as alterations in the levels of DNA methyltransferase (DNMT), DNA methylation as well as histone acetylation and methylation patterns adversely affect the oocytes [5, 6]. Put together, epigenetic modifications could affect gametogenesis as well as the embryo development since they involve germ cells and can therefore be transmitted to the offspring [7]. Since aging has been linked with epigenetic and reproductive alterations as well as increase in oxidative stress, which results in profound biological consequences, this chapter shall discuss the interconnection among epigenetic alterations (histone modification, chromatin remodeling, DNA methylation and miRNA), reproductive aging, and oxidative stress.

2. Reproductive aging

Aging, an unavoidable and unappeasable natural process, has markedly disproportional gender-based impressions on human fertility. As human age, their reproductive functions diminish. Both male and female sex hormones diminish with age. While the male sex hormones (androgens) and their breakdown products diminish gradually over the age span 50–90, the female sex hormones (estrogens) fall significantly at menopause. Moreover, in both males and females, sexual activity is reduced progressively between the ages of 20 and 60 [8, 9].

In the practical sense, all males between the age of 20–45 have some level of sexual activity. However, the frequency of intercourse in males may fall from an average of four per week in 20-year-olds to one per week in 60-year-old. This means that there is loss of sexual activity in males in only about 5% of the ages between 45 and 60. Moreover, there are certain systematic studies with reports of sexual behavior in individuals above 60 years of age even though at least some males remain sexually active at age 90 according to clinical reports. By comparison of sexual activity in both males and females, there are wide individual differences in the level of sexual state of being active and sexual behavior, which is greatly influenced by psychological and social factors than by the levels of sex hormones circulating in the blood. Notwithstanding, the use of male sex hormones has been known with males since time immemorial [10].

The reproductive functions in males (both spermatogenesis and testosterone production), diminish with age even though slowly and to a relatively small degree. Changes associated with aging in the reproductive system of males may include changes in production of sperm, erection, and testicular tissue, all of which are gradual [11]. In human, males do not experience a major, rapid (over several months) change in fertility as they age unlike like menopause in females. The changes in males are associated with a steady and gradual process regarded as andropause. The testes are the primary organ in the male reproductive system where the change in aging takes place. The testes are the organ saddled with the function of making the male sex hormones (testosterone), the level which decreases with aging during which erection problem may arise. This process is however, a typically slow rather than a sudden and complete lack of function [12]. With aging, the testes can actually continue to make sperm but at a much slower rate of producing viable sperm cells. Also, other accessory organs of male reproduction function at a reduced rate; the epididymis, seminal vesicles, and prostate gland will continue to produce the fluid for easy sperm movement even though they lose some of their surface cells with aging. When all these occur, a condition known as sclerosis sets in such that the tubes that carry sperm may become less elastic. While a fall in testosterone level may be associated with primary hypogonadism in some men, it may be linked with secondary hypogonadism, accompanied by illness in other men [11]. Basically, a fall in the production of testosterone is concomitant with certain consequences such as diminished sexual function, mood, energy, bone density, and muscle mass. Therefore, energy or cognitive function are not improved by increasing the serum testosterone of older men with low testosterone to that of young men even though bone density, sexual function, mood, walking, and ultimately sexual function are well improved [13, 14].

In females, they often experience sudden loss of fertility between 30 and 40 years of their age. Within this age, females have a significant increase in the chances of having chromosomal defects in offspring, preterm delivery, spontaneous abortions, and intrauterine growth restriction (IUGR). This is accompanied by the onset of menopause [15, 16]. Contrastingly, this is not the case in males where the impact of their aging has not received much attention like that of female yet have the more striking negative consequences on the process of aging [17].

Reproductive aging in females ends in menopause as a result of natural processes in the hypothalamic–pituitary–ovarian (HPO) axis. At menopause, the HPO axis is in a state of hypergonadotropic hypogonadism. This reflects a minimal ovarian estrogen production and an accompanying increase in pituitary gonadotropins. Most women become aware of reproductive aging by alterations in menstrual cyclicity or local/systemic symptoms of hypoestrogenism after their final menstrual period between age 49–51. At menopause, changes in the female reproductive organs occur rapidly and when it stops, and the ovaries stop producing estrogen [18, 19].

At menopause, changes in the female reproductive organs occur rapidly and when it stops, and the ovaries stop producing estrogen. Moreover, atrophy sets in after menopause such that the tissues of the labia minora, clitoris, vagina, and urethra become thin. A resultant effect of such thinning is chronic irritation and dryness of the vagina [20]. In addition, there is a more likely chance of women at this stage to develop urinary tract infections and vaginal discharge apart from the fact that the ovaries, uterus, and fallopian tubes. There is also decrease in the amount of muscle and connective tissue (including that in muscles, ligaments, and other tissues that support the bladder, uterus, vagina, and rectum) with aging.

More importantly, changes in reproductive organs associated with age do not interfere with sexual pleasure even though the dryness of the vaginal after menopause causes pain during sexual activity while there might also be a drop in the desire to have sex in some women [21, 22].

3. Epigenetic changes associated with reproductive aging

Epigenome is an all-important and targets capable of being modified. They include the methylation of DNA, nuclear protein constituents (and detailed modifications to protein tails of histone) and may also include various RNA species. These epigenetic targets have the ability to regulate the expression of gene and being able to be passed onto the embryo proceeding fertilization. The epigenetic profile of cells in the body is unique specific attributed function owing to the role of epigenomes in the regulation of genes. In humans, sperm and egg cells also have these features and this is the basis of the uniqueness of cell type with a highly specialized epigenome well suited for morphologically and functionally distinct attributes [23].

Epigenetics can be described as heritable covalent modifications of the DNA bases and chromatin proteins whereby the DNA sequence is not altered but regulate its transcriptional process by influencing the chromatin structure and transcription factor binding [24]. Epigenetics are vital to the usual development and functioning, and as well as participate as an essential constituent or characteristic in both normal cellular function and disease. Generally, epigenetics changes with age as part of the normal human development or aging, and as a reaction to human behaviors and environment. The three major epigenetic modifications include DNA methylation, histone modification, and chromatin remodeling [25].

As far as human development is concerned, epigenetic changes begin before birth since all cells have the same genes but phenotypically different. Epigenetics determines the functional role of each cell as either a heart cell, nerve cell, or skin cell with growth and development. For example, the nerve cells and muscle cells may have the same DNA, they exhibit different functional roles. While a nerve cell transfers information to other cells, a muscle cell has a structurally helps the body movement. In this regard, epigenetics modulates the muscle cell to turn "on" genes to make proteins important for its job and turn "off" genes important for a nerve cell's function [25].

Also with age, epigenetics changes throughout the reproductive life in human. The epigenetics at birth is different from at childhood or adulthood. For instance, in a study that compared the epigenetics of an infant, 26-year-old and 103-year-old using DNA methylation at millions of sites, it was reported that the level of DNA methylation decreased with age. The infant had the most eminent DNA methylation while the 103-year-old had the most depleted DNA methylation. The DNA methylation level of the 26-year-old was an intermediate of the newborn and 103-year-old [26]. In addition, epigenetics is reversible. There are epigenetic changes that may be gotten rid of or added as a result of behavior or environment changes, an indication that epigenetic changes can be temporal. In a comparative among a smokers, non-smokers, and former smokers, it was found that smoking can result in epigenetic changes. DNA methylation at specific parts of the *AHRR* gene tend to be lower in smokers compared to non-smokers, while there was a wide difference for heavy smokers and long-term smokers. After quitting smoking, former smokers had increased DNA methylation at this gene and eventually reached similar levels as those of non-smokers [27].

Epigenetic alterations symbolize one of the distinctive characteristics of aging. It is a significant and crucial mechanism associated with the degenerated cellular activities and functions during aging. Epigenetics therefore explains why aging pattern vary between two identical twins. Genomic instability transcriptional drift occur as a result of unevenness in the pattern of epigenetic information within individual cells in the population during aging (**Figure 1**) [28].

Furthermore, the information encoded within different epigenome includes DNA methylation, chromatin remodeling, posttranslational modifications of the histone proteins, structural and functional variants of histones, and transcription of non-coding RNAs (ncRNAs). The combination of all these different types of epigenetic information comprises the function and fate of all cells and tissues [29].

DNA methylation (one of the best-characterized and most widely studied epigenetic modifications during aging) involves adding a methyl group, primarily from S-adenosylmethionine to the fifth carbon position of cytosine bases in the CpG dinucleotide regions [30]. The DNA methyltransferase group of enzymes catalyzes DNA methylation. DNA methylation represses the transcription of a gene by condensing the chromatin structure, preventing transcription factors from binding and decreasing the process of histone acetylation, whereas DNA methylation of the 3' CpG islands activates the transcription of genes. Epigenetic

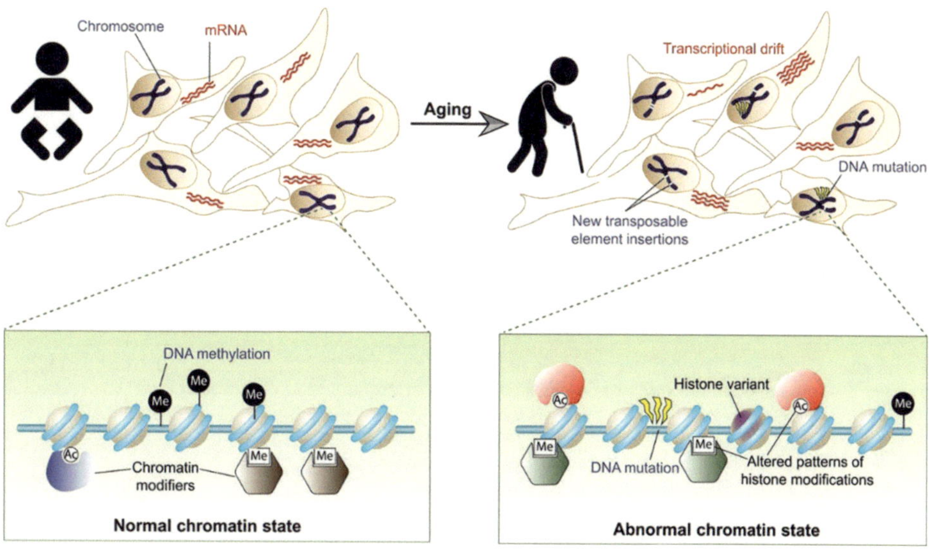

Figure 1.
Summary of epigenetic changes associated with reproductive aging. The cells within each cell type in young persons have an almost the same pattern of gene expression, mostly determined by each cell with related epigenetic information. However, these epigenetic information changes in a sporadic manner as one age because of endogenous and exogenous factors. The resultant effect is abnormal chromatin state with the uniqueness of altered patterns of DNA methylation, incorporated histone variants of different origins, and altered patterns of histone modification, which leads to different chromatin modifiers. Insertion of new transposable elements into the genome and genomic instability (including DNA mutation) is also a resultant effect of abnormal chromatin state in old cells.

drift results to irregular divergences in the methylome among aging individuals. Aging in humans as in all other mammals is associated with hypomethylation of CpG (although with a few exceptions) especially at repetitive DNA sequences. As a whole, age-related DNA methylation changes are more prominent in CpG islands (**Figure 2**) [31].

In addition, epigenetic modifications involves the complexity of epigenetic processes, associated with DNA modifications—5-methylcytosine (5mC) DNA methylation, 5-hydroxymethylcytosine (5hmC) DNA methylation; Histone modifications—Acetylation, Methylation, Phosphorylation, Poly-ADP ribosylation, Ubiquitination; Non-coding RNA interactions—piwi RNA (piRNA), small interfering RNA (siRNA), long non-coding RNA (lncRNA), micro RNA (miRNA); RNA modifications—6-methyladenosine (6 mA) RNA methylation, 5-methylcytosine (5mC) RNA methylation, 7-methylguanosine (7mG) RNA methylation, mRNA CAP, 5-hydroxymethylcytosine (5hmC) RNA methylation (**Figure 3**).

Histone modification occurs in varying forms: acetylation, methylation, phosphorylation, ubiquination, and sumoylation. These forms either repress or activate transcription based on the location of histone modification [32, 33]. Chromatin remodeling is the rearrangement of chromatin from the condensed heterochromatin state to the relaxed euchromatin state, allowing transcription factors or other DNA-binding proteins to access DNA and regulate gene expression. Chromatin remodeling is highly implicated in epigenetics and generally results in transcription activation [34]. Non-histone protein methylation forms another crucial cellular function regulator as post-translational modifications that takes place on proteins and have the ability alter their function. The modifications allow for the addition of methyl groups to lysine or arginine residues of specific proteins. Even though the functional role of

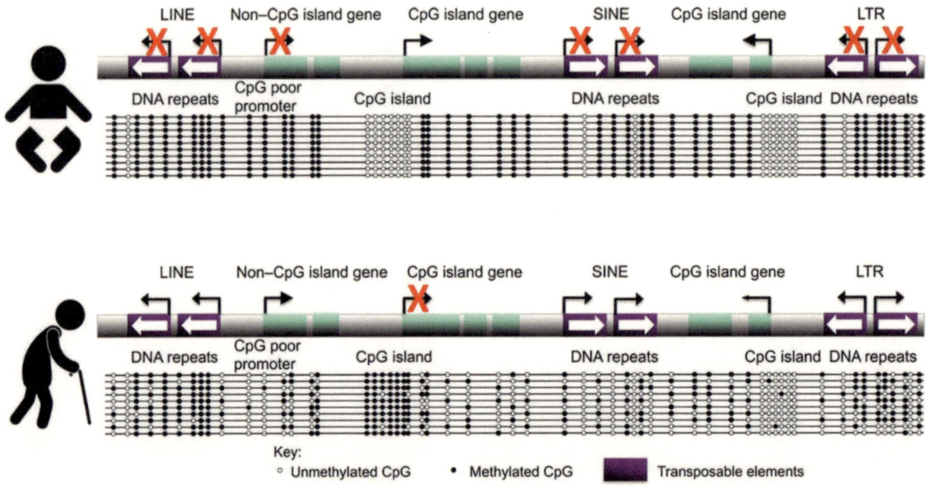

Figure 2.
Summary of changes in DNA methylation during reproductive aging. Young mammalian cells are distinguishable by hypermethylation of DNA over the genome, except for CpG islands within the promoters of expressed genes. DNA repeats (like as LINE, SINE, and long terminal repeat transposable elements) are DNA-methylated to a great extent which aids their upkeep in a state of constitutive heterochromatin. Hypomethylation of DNA (within the cell population in a stochastic manner) generally occur over the genome during aging but the loss of DNA methylation results to the activation of the transposable elements as an example of other normally silenced DNA sequences. Also in a stochastic manner, methylation of DNA increases over the CpG islands of certain genes in relation with their silencing and heterochromatinization.

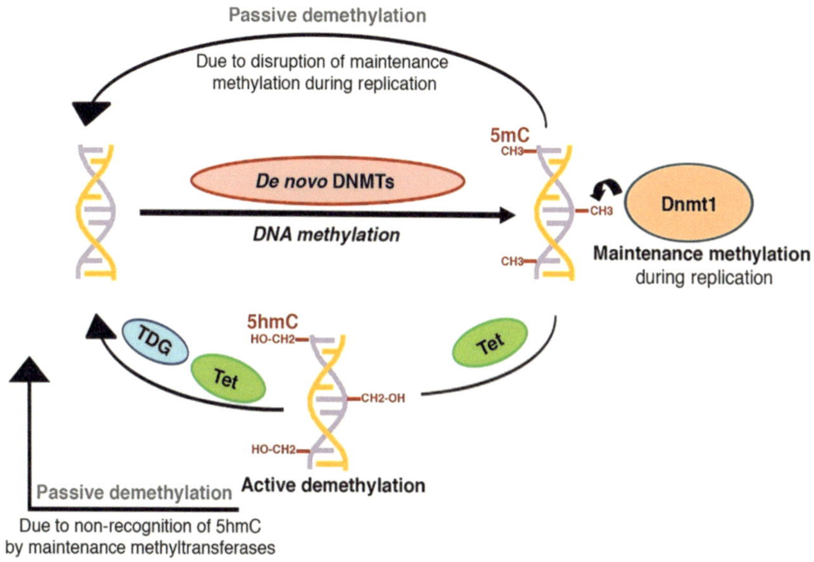

Figure 3.
Schematic illustration of DNA methylation patterning. During development, de novo DNA methyltransferases regulates the formation of new DNA methylation patterns. During regular succession of cell division, DNA methylation patterning is made to continue by activity of maintenance DNA methyltransferases. However, active demethylation [successive enzymatic oxidation of 5-methylcytosine (5mC) to 5-hydroxymethylcytosine (5hmC),5-formylcytosine (5fC), and 5-carboxylcytosine (5caC) by TET (ten-eleven translocation) dioxygenases, followed by thymine DNA glycosylase (TDG) dependent removal of 5fC and 5caC, coupled with base-excision repair to a cytosine (C)] or passive demethylation could reverse DNA methylation marks. A passive demethylation ensues when there an unrecognized hemi-methylated 5hmC by the maintenance DNA methyltransferases because of being diluted and lost during replication. Disruption of maintenance methyltransferase activity can similarly result in replication dependent dilution of DNA methylation.

these modifications is poorly understood, they are more likely to affect the structure and function of DNA, synthesis and metabolism of protein and RNA, as well as apoptosis and cell cycle [35].

3.1 Epigenetic changes associated with female reproductive aging

Epigenetic changes and epigenetic related enzymes in the oocytes of aged females are found to have altered DNA methylation and DNA methyltransferase levels. They also have altered patterns of methylation and acetylation of histone [6].

DNA methylation is catalyzed by the DNA methyltransferase group of enzymes. In mammals, there are five known forms of the enzyme viz.: DNMT 1, DNMT 2, DNMT 3a, DNMT 3b, and DNMT 3 L [6]. DNA methylation occurs either as maintenance or *de novo*. In methylation maintenance, a hemimethylated DNA becomes fully methylated. This process occurs after semi-conservative DNA replication and is carried out by DNMT1 [36]. In *de novo* methylation, a nascent and completely unmethylated double-stranded DNA is methylated. This process is the sole responsibility of DNMT 3a and DNMT 3b. DNMT 3 L does not methylate DNA but facilitates the activity of DNMT 3a and DNMT 3b, while DNMT 2 is separately involved in the methylation of transfer RNA [37].

The levels of DNMT 3a, 3b, and 3 L in developing oocytes have been shown to correlate with their levels of growth and DNA methylation, associating them with a unique role in oocyte development [6]. The pattern of DNMT regulation is altered with aging, and a decrease in DNA methylation due to decreased level of DNMT transcription is observed in aging oocytes [38].

The demethylation and remethylation patterns of the DNA observed during oocyte development are altered with maternal aging. These alterations have been associated with decreased expression of many important genes. With increasing maternal age, expression was decreased in over 800 genes including many genes that are critical for cell cycle control and meiotic chromosomal segregation and are potential causes of aneuploidy [39].

3.1.1 Alterations of DNA methyltransferase during reproductive aging in females

The class of enzymes involved to add methyl groups to DNA is regarded as DNA methyltransferase (DNMT). DNMT1, DNMT2, DNMT3a, DNMT3b, and DNMT3L are the five types of DNA methyltransferase known with mammals while *de novo* and maintenance methylation are the two types of DNA methylation [40]. There is uncertainty in the levels of the various types of DNMT when an oocyte develops during normal developmental stages from primordial to primary, to secondary follicle, and then from the germinal vesicle (GV) through MII and beyond. The expression of DNMT1 is first noticed in the secondary follicle stage and then at the zygotic stage and beyond [6]. In the case of DNMT3a, the expression is right from the primordial stage while the expression of DNMT3b is from the primary follicle stage. Furthermore, while DNMT3L is found in pre-implantation embryos, DNMT2a is not observed at any stage. Changes are associated with cellular location of the enzymes as either cytoplasmic or nucleic at the different stages of development. DNA methylation and the levels of growth of the developing oocytes have correlations with the levels DNMT 3a, 3b, and 3 L in the oocytes, which is associated with the maturation of the oocytes [38, 41]. Similarly, embryonic death as a result of imprinting failure (epigenetic silencing of either maternal or paternal DNA linked to the expression of just one chromosome for specific trait) associated with lack of methylation are the resultant effects of targeted gene deletions of DNMT3a or DNMT3L. This suggests the crucial function of DNMT3a or DNMT3L as demonstrated in the development of oocytes such that oocytes without DNMT3b do not exhibit grave abnormality [42].

As a female age, the regulation pattern of DNMT is altered. In experimental study on young mice of about 6 weeks old and aged mice of about 45 weeks old using MII oocytes, it was found that there was altered levels of genes transcription (including DNMT1, DNMT3L, and higher levels of DNMT3b transcription) associated with setting up and preserving DNA methylation [6]. In a similar manner, in a study that used 66 weeks old older mice, it was found that there was diminished level of transcription of DNMT3a. Furthermore, the human oocytes from aged females in another study revealed a decreased transcription of the genes associated with the cell cycle checkpoint as well as transcription and the DNA damage repair [6, 43].

Furthermore, it was found that there was a noticeable and obvious decrease in the aged mice of 40 weeks relative to the young ones of about 8 weeks old in the attempt to compare pre-implantation embryos (2 cell, 4 cell, 8 cell, and morula) and levels of DNMT1, DNMT3a, DNMT3b, and DNMT3L in MII oocytes. Such decrease can be linked to a wider reduction in DNA methylation of aging oocytes [38].

3.1.2 Alterations in methylation of DNA during reproductive aging in females

Dynamism is associated with DNA methylation in both early embryos and in germ cells. Methylation of DNA is crucial in aging. The stability and highly methylated

DNA associated with somatic cells afford them the regulations of expression their genes for easy functions specific to tissues. Also, oocytes and sperm exhibit high methylation levels as they go through constant changes all through their development state [44]. At the initial primordial stage, mouse germ cells undergo genome wide demethylation. Oocytes then enter meiotic arrest, and undergo remethylation only after birth, during oocyte growth from primary to secondary follicles. Male germ cells replicate throughout the life of the male and enter meiosis in the adult male. Remethylation in male germ cells occurs antepartum at the pro-spermatagonia stage [45]. Both the maternal and paternal chromosomes are physically separate at post fertilization whereby they undergo varying methylation changes. As for the paternal genome, it is actively-demethylated prior to DNA replication, while a passive demethylation occurs with the maternal genome. As at the blastocyst stage close to the time of implantation, both genomes are again remethylated. This cycle of demethylation and remethylation is crucial for the removal of parental epigenetic modifications to the germ cell genome, and for setting up totipotency in the new embryo [6, 44].

As a female age, these patterns of alterations change. As previously described, it has been demonstrated that level of DNA methylation in older mice of about 40 weeks old were lower than those younger mice of about 8 weeks in a study that compared pre-implantation embryos (2 cell, 4 cell, 8 cell, and morula) and 5-MeC fluorescence intensity in MII oocytes [6, 38]. However, DNA methylation did not produce any noticeable change in the blastocysts and the whole observation was linked to *de novo* methylation that takes place through DNMT3a and DNMT3b before the implantation happens. Furthermore, there were indications that with aging, the usual attainment and preservation of DNA methylation patterning is associated with the capability of embryo to develop to mid-gestation [46, 47]. DNA demethylation is what is achieved as the opposite of DNA methylation is DNA demethylation. DNA demethylation has been found to increase with age, a direct opposite of low levels of DNA methylation associated with old age. This demethylation cascade demonstrated using mouse oocytes can be linked with decreased methylation and increased demethylation [48]. In addition to this, chemically induced accelerated aging as been shown to demonstrate a varying category of demethylation-pathway-intermediates relative to the normal process aging. This may find usefulness in differentiating natural aging from accelerated aging, and importantly help to evaluate the rate of reproductive aging in women to achieve an estimated reproductive longevity [6, 49].

The consequential effect of alteration of DNA methylation in aged females is associated with decreased expression of several crucial genes. For instance, decrease in more than 800 genes (including several important genes for the meiotic chromosomal segregation and control of cell cycle) were found in the aged, analyzed for genetic expression of human blastocysts using single embryo RNA-seq. The crucial role epigenetics plays in normal reproduction was described in all these [39].

3.1.3 Modifications of histone during reproductive aging in females

Acetylation of the N-terminal of histones takes place on lysine (K) residues and promotes transcription. Methylation of histone occurs on lysine or arginine residues. Depending on location, histone modification function to either help in transcription suppression or promotion. While the transcription suppression is linked with the methylation of histone H3 K9, transcription promotion is associated with methylation of H3 K4, or methylation of arginine on H3 or H4. Modifications of histone are crucial to normal gametogenesis and just like the methylation of DNA, they are in constant change during the development of germ cells. Similarity was found in the acetylation of histone patterns in porcine, bovine, and sheep although slightly different pattern were noticed. However, there were relatively stable during oocyte maturation in histone methylation [31].

As a female age just with alteration in DNA methylation changes, the patterns of modification also shift. It was found that the reduced levels of acetylation at H4K12 and H4K16 in aged mice of about 10 months old relative to younger mice of about 2 months old using the germinal vesicle oocytes. In the younger mice, 100% was recorded for both H4K12 and H4K16 as against 67/81 for H4K12 and 55/92 in the H4K16 of the aged mice [6, 43]. Furthermore, 40% of oocytes from older mice were acetylated at H4K12 as against the 100% complete deacetylation at H4K12 of all oocytes from young mice. Moreover, while using an inhibitor of histone deacetylase, Trichostatin A (TSA), corrections were made to errors in acetylation of aging related MII stage related to acetylation of germinal vesicle oocytes [50]. In a similar manner, another evidence was provided in a comparison study of MII oocytes from the same mouse to get rid of genetic interactions. In the 10 months old mice, found higher H4K12 acetylation levels was recorded relative to the 3 weeks old young mice. This further shows the clinical usefulness of H4K12 acetylation levels as a biomarker for oocyte quality [51, 52]. These changes in histone modification results in oocyte dysfunction and infertility. An increased levels of histone acetylation in MII mouse oocytes described the associated between inhibiting histone deacetylase during meiosis, and a high frequency of oocyte aneuploidy and embryo death [53].

The age of oocyte has also been found to correlate with unusual high histone acetylation levels on H4K12 in the MI and MII of human oocytes. This suggests a greater tendency of misalignment among chromosome which may lead to more segregation errors in older oocytes and by implication, becomes a useful clinical tool. In a similar vein, histone modification patterns, particularly histone acetylation and methylation, are markedly altered with maternal aging and can lead to oocyte dysfunction and infertility [54]. The alterations that occur during histone phosphorylation, ubiquination, and sumoylation in the oocytes of older females are still poorly understood but dysregulation in histone ubiquination in aged oocytes has been observed suggesting oocyte dysfunction in aged females [55].

3.2 Epigenetic changes associated with reproductive aging in males

Epigenetic alterations have been associated with the sperm of aged males and are demonstrated to impact male fertility, embryogenesis, and even offspring health [56]. Oakes *et al.* observed that age-associated alterations in DNA methylation occur at specific genomic loci of the sperm of male rats. Other research efforts have also brought to light the profound alterations in DNA methylation that occur in aged spermatozoa [57–59]. However, a consensus on the mechanism of alterations in DNA methylation has not been reached due to the varying experimental approaches utilized by the researchers.

Very limited studies are available on the alterations in histone modification patterns with respect to age. Indeed, only two studies are available where histone modifications in spermatozoa are analyzed in relation to age and certain alterations were observed in the histone modification patterns between young and aged mice [60, 61].

3.2.1 Involvement of epigenetics in spermatogenesis

Mitotic proliferation of spermatogonia, meiotic divisions, and morphological differentiation of sperm precursors (spermiogenesis) are the processes involved to form a matured sperm. These processes are also associated with specialization of cells distinguished by the presence of a head, an intermediate portion, and a flagellum [62, 63]. Such a specific system of arrangement of the male germ cells allows the movement of sperms through a potentially uncongenial female reproductive tract, penetrate the cumulus oophorus and the zona pellucida, penetrate the oocyte, and eventually make a complete multiple post-penetration [64, 65]. Spermatogenesis is initiated in the seminiferous tubules during fetal development from a spermatogonia (undifferentiated diploid cells) which go through some stages of mitotic divisions to make germ cells precursors more available [66]. At maturity in males, there are about three stages of transformation of spermatogonia. During the first meiotic division, some are transformed in type I spermatocytes that produces haploid type II spermatocytes while haploid spermatids are formed during the second meiotic division. The last and third stage portrayed by structural and morphological transformation complex action of the round spermatid is symbolized by spermiogenesis. This last stage of spermatogenesis which ends subsequent cell division forms the mature sperm with definite characteristics of differentiated flagellum and acrosome that forms a vital requirement for motility of the produced sperm as well as the fertilization capacity [67].

3.2.2 Main epigenetic change in sperms: histone: protamine substitution

A well-arranged and coordinated chromatin structure is essential to characterize sperms asides their distinctive morphology and motility. During spermiogenesis sperm chromatin are further condensed due to the substitution of about 95% of the histones with protamines (sperm-specific basic proteins) which leads to disulfide bonds (SS) formation [68]. These bonds give core of the sperm nucleus a high degree of stability responsible for desirable sperm motility, shielding from ROS and toxicants within female reproductive tract, blockage of the transcriptional activity of the sperm DNA, and other notable effects to the sperm [69]. This cascade of reaction of multi-step process of conversion of histone to protamine is highly regulated. At first step regarded as histone hyperacetylation, there is the replacement of the histones in round spermatids by transition proteins (TP, heterogeneous group of nuclear proteins). The second stage has to do with the substitution of TP1 and TP2 with protamines and takes place in elongating spermatids. Protamines in this regard function to ensure the genetic wholeness of the sperm and epigenetic imprinting via making the nucleus more compact. Two types of protamines (the P1 protamine and the P2 family of protamines, made up of P2 as the most abundant, P3, and P4 members) are notably known with nuclei of a matured spermatid [70]. The ratio of P1P2 to play a crucial role in male fertility. For instance, P1/P2 ratio, which in fertile males is close to 1 (range 0.8–1.2), is altered in infertile patients. Therefore, patients having a P1/P2 ratio less than 0.8 demonstrate insufficient DNA condensation and the characteristics of sperm (such as motility, viability, vitality, and counts, and morphology) are resultantly altered. A lower P1/P2 ratios may also be linked with an increased DNA fragmentation, which in contrary manner have correlations with the levels of global sperm P1 and P2 [71, 72]. This confers a protective

S/N	Model(s)	Markers	Epigenetic change noted with reproductive aging
1	*Caenorhabditis elegans*	Reduction in HP1 and H3K9me3, changed lamin A	Reduced global heterochromatin
	Drosophila		
	Human fibroblasts		
	Progeria patient cells		
2	Mice	Increase in HP1 and H3K9me3, as well as increase in macroH2A and HMGA	Senescence-associated heterochromatin foci (SAHF)
	Baboons		
	Human fibroblasts		
3	Yeast	Loss of core histone proteins	Remodeling and loss of nucleosome
	Worms		
	Human fibroblasts		
4	Mouse brain	H3.3, H3.3cs1	Increased histone variants
	Human fibroblasts		
5	Yeast	Increase in H3K4me3 and H4K16ac (globally active marks) and decrease in H3K9me3 and H3K27me3 (repressive marks)	Changed histone marks
	Drosophila		
	C. elegans		
	Killifish		
	Rats		
	Mouse brain		
	Mouse stem cells		
	Mouse fibroblasts		
	Progeria mouse models		
	Progeria patient cells		
	Human fibroblasts		
	Human brain tissue		
6	Salmon	Hypomethylation of lobal DNA, hypermethylation of CpG island	Changes in DNA methylation
	Mice		
	Rats		
	Dogs		
	Rhesus monkeys		
	Human fibroblasts		
	Human stem cells		
	Humans		
7	Yeast	SIRT-1, PARP-1, REST, HDAC-1	Re-localization of chromatin-modifying factors (RCM)
	Mice		

S/N	Model(s)	Markers	Epigenetic change noted with reproductive aging
8	Yeast	H19, Dicer, lin-4, lin-14, mir-34	Changes in ncRNA
	C. elegans		
	Mice		
	AD mouse models		
	Humans		

Table 1.
Epigenetics changed genes in reproductive aging.

role on protamines against sperm DNA damage. Furthermore, a surplus protamine P2 precursors (pre-P2) may be linked with subfertility which alters the process of formation of mature protamine P2 [73].

3.2.3 Methylation of DNA and modifications of histone during spermatogenesis

For thoroughly maturation of gametes in male, there are need for several and specific epigenetic marks during gametogenesis. For instance, demethylation of DNA can be said to be the initial epigenetic occurrences prior to meiosis but during meiosis, the de novo DNA methylation levels are regulated by the activity of DNMT3A, DNMT3B, and cofactor DNMT3L, to complete this process after birth at the pachytene spermatocyte stage [44, 74]. Thereafter, DNMT1 activity maintains methylation profile during which the modification of histone (methylation and acetylation) takes place. This modification eventually alters the DNA accessibility to transcription factors. Most importantly, histone demethylase (HDM) and histone methyltransferase (HMT), specific for this action regulate methylation sequence of histone H3 (H3-K4) and lysine 9 of histone H3 (H3-K9) [31]. In a typical setting, methylation of histone H3-K9 which is eminent known in meiosis is absent at the termination of the process for the promotion of genes. On the other hand, methylation of histone H3-K4 is greatly reduced during meiosis for DNA silencing. Furthermore, HAT and HDAC and other enzymes are involved on the regulation of H3 and H4 lysine residues acetylation and deacetylation during spermatogenesis. In this process of spermiogenesis hyperacetylation of H4 is crucial to make the right conversion of histone to protamine and as well the disassembling of nucleosome seamless in elongating spermatids [31]. **Table 1** gives summary of epigenetics changed genes in reproductive aging [75].

4. Interconnection among oxidative stress, reproductive aging, and epigenetics

4.1 Link between reproductive aging and oxidative stress

Oxidative stress results from an imbalance between the production of reactive oxygen species (ROS) in cells and tissues and their rate of elimination or detoxification. Typically, oxidative phosphorylation in the mitochondria produces ROS as a byproduct. For the body to operate normally, there must be a moderate quantity of

ROS. ROS have a critical role as a second messenger in intercellular signal transduction, the control of gene expression, and immunological function in addition to their roles in the manufacture of active chemicals, cellular detoxification, and immune function [1]. But when ROS are overproduced or antioxidant use rises, redox processes become unbalanced and the body enters an oxidative stress state [76].

The process of aging in the ovaries exacerbates the production of reactive oxygen species. In particular, age-related ovarian aging causes the accumulation of ROS in the oocytes [77]. Age-related oocyte aging has been associated with a lowered expression of key genes in the ETC of oocytes [78]. For example, the succinate dehydrogenase complex flavoprotein subunit A (Sdha) and coenzyme Q10 (CoQ10) proteins that are responsible for shuttling electrons through the electron transport chain, have an age-related down regulation in mouse oocytes [79]. The dysfunctional shuttling of the electrons through the electron transport chain results in electron leakage from the mitochondria. The leaked electrons initiate a chain reaction where electrons are extracted from lipids, predominately polyunsaturated fatty acids including arachidonic acid and linoleic acid, and ultimately results in lipid peroxidation. Lipid peroxidation produces numerous lipid aldehyde byproducts that have the potential of damaging mitochondrial DNA/RNA and proteins. In addition, age-related oocyte aging has been linked to the accumulation of advanced glycation end-products (AGEs). AGEs increase ROS generation by the induction of hypoxia through collagen crosslinking and the impairment of perifollicular vascularization [80]. Moreover, the decreased antioxidant capacity in age-related ovarian aging is another cause of ROS generation in oocytes [78, 81].

Similarly, oxidative stress has been associated with age-related DNA sperm damage. In fact, the majority of DNA sperm damage is a result of oxidative stress, as evidenced by high correlations observed between the production of the major oxidative adduct 8-hydroxy-2′ deoxyguanosine (8-OHdG) and DNA fragmentation in sperm cells [82]. DNA sperm damage reduces male fertility and leads to inappropriate oocyte fertilization (**Figure 4**). This type of fertilization is associated with maternal miscarriages, preterm births, genetic diseases, neurological disorders, and juvenile cancer [83]. Furthermore, age-induced oxidative stress can result in the loss of membrane fluidity in sperm cells, consequently reducing sperm motility [3].

Several theories have been raised as a result of the consequential effect of oxidative stress in the biological system especially in aging. One of such theories regarded as the "free radical theory of aging" posit that aging is a resultant deleterious effects of free radicals accumulated overtime. Meanwhile in agreement with this postulation, mitochondria production of ROS (proposed as the key causative factor of aging) has been found to increase in aged tissues [84]. There are more scientific evidence to support this theory whereby increased oxidative damages in cells are associated with aging. The resultant effects of this were reported to be cellular dysfunction from accumulated damages in nucleic acids, lipids, carbohydrates, and proteins that exposes the body to harmful attacks of external agents easily. Furthermore, 8-oxo-7,8-dihydro-2′-deoxyguanosine (8-oxo-dG) are found in mtDNA in a more eminent concentration compared to the nuclear DNA [85]. This indicates that mtDNA is highly vulnerable to oxidative damage; moreover, the mitochondria of mammalian cells are major producer of ROS. This further improve on the previous theory on free radical regarded as the mitochondrial theory of aging. This latter theory posits that oxidative phosphorylation of mitochondrial macromolecules (including lipids, proteins, and mtDNA) generates oxidative damages that result to aging [86]. In addition, since the regulation of apoptosis is majorly modulated in the mitochondria, it is suggested that

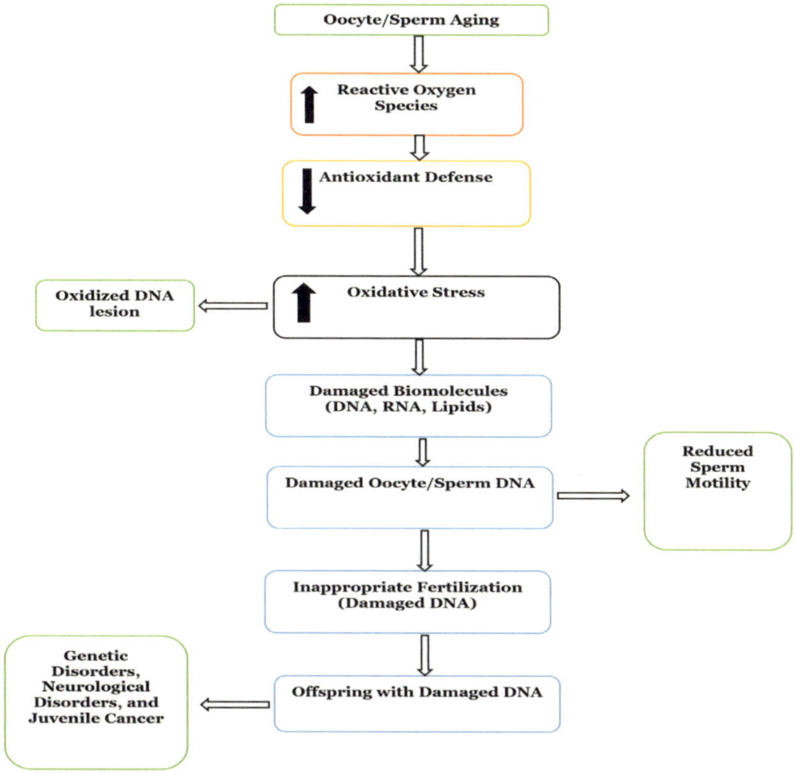

Figure 4.
Consequences of oocyte/sperm aging. Oocyte and sperm aging results in the production of reactive oxygen species through electron leakage from the mitochondria. Also, it causes a reduction in the cellular antioxidant system. The combination of these factors causes oxidative stress, which can directly result in oxidized DNA lesions or indirectly cause damaged biomolecules (DNA, RNA, lipids). In particular, the oocyte/sperm DNA is damaged, resulting in reduced sperm motility in aged males and inappropriate fertilization during copulation. Upon birth, offspring with damaged DNA exacerbates the possibility of genetic disorders, neurological disorders, and juvenile cancer.

aging due to apoptosis are associated with age-related mitochondrial oxidative stress. Asides, the beneficial outcome of antioxidants targeted at the mitochondria has been reported by several studies, which are indications that targeted antioxidants protect against oxidative damage better than the untargeted cellular antioxidants in the mitochondria [85, 87]. One suitable reason for this is ability of the targeted antioxidants to get rid of ROS at specific points for being able to cross the mitochondrial phospholipid bilayer [88].

4.2 Interplay between oxidative stress and epigenetics

Oxidative stress is a resultant effect of accumulated ROS which increases with age. It is also associated with a declination of the cells' repair machinery that consequentially generate a broad scope of DNA lesions [89]. This cascade of events results to mutated genes and a disrupted epigenetic state of the cell. Several studies have suggested the interplay that exists between epigenetics and oxidative stress viewpoint. For instance, demethylation of DNA by the oxidation of DNA and hydroxymethylation (mediated by TET) can be influenced by ROS. Also, ROS affects the methylome via oxidized DNA lesions formation [a resultant effect of hydroxylation of 5-methylcytosine (5mC) and pyrimidines]. This hydroxylation reaction however may be a hinderance owing to the similarity in structure with epigenetic signals associated with 5-hmC [90, 91]. Furthermore, essential metabolites such as S-adenosylmethionine, acetyl-CoA, ketoglutarate, NAD^+, and Fe modulates histone-modifying enzymes that participates in epigenetic machinery [92]. However, activity of the epigenetic machinery can be influenced by ROS which suggests that epigenetic changes are associated with energy capacity of the cell and entire cellular metabolism. Oxidative stress, therefore, have a great impact on the epigenetic expanse of the cells at various control point ranging from DNA and histones to histone modifiers [93].

4.3 Oxidative stress and epigenetic deregulation

Oxidative stress from several factors causes epigenetic modifications. Moreover, molecular mechanisms associated with aging are involved in processes like the methylation of DNA, noncoding RNA, and histone change, generally regarded as epigenetic modifications [90]. In males, the loss DNA methyltransferases results to an incomplete formation and maturation of gamete. Specifically, impairment of sperm DNA methylation occur with oxidative stress on the cytosine-guanine (CpG) islands. Moreover, 8-oxodeoxyguanosine is formed from oxidation of guanine while 5-OH C, 5,6-diOH C, C glycol are formed from alteration of cytosine whereby 5-hydroxymethylcytosine formed as physiological product of the oxidation is the needed for the demethylation of DNA [5]. Also, the deamination of 8-oxoG, 5-MeC, or 5-MeC to 5HmC leads to the formation of thymine which hinders the methyl CpG-binding domain proteins to be bonded to matching CpG. This results to a poor methylation because of a weak binding of the DNA for DNMT3A [5, 94]. By consequence, oxidative damage to DNA generates changes in heritable epigenetics via alterations to the chromatin arrangement. An uncontrolled concentration of 5HmC (the oxidative product of cytosine) also results in the stimulation the process of active aberrant demethylation of DNA. Loss of KMT2D (Lysine methyltranseferase 2D also known as H3K3 monomethyltransferase) expression is also presumed to be associated with DNA damages mediated by ROS. Loss of KMT2D expression results to a reduction in the number of enhancers like H3K4me1 and H3K27ac regarded as enhancer activity markers. This reduction in their abundance hinders the binding of transcription factor like the FOXO3 and resultantly confer protection against oxidative stress through the upregulation of enzymes like superoxide dismutase and catalase and other antioxidant enzymes [95]. Furthermore, suppressed action of KMT2D is associated with DNA damage via the accumulation of ROS in prostate cancer condition. For instance, it was shown that there is a connection in lower sperm count and DNA methylation in men that has altered spermatogenesis in H19 and PEG1/MEST regions [96]. Similarly, hypomethylation induced by ROS is also associated with del Castillo syndrome or germ cell aplasia, cancer of the testes, and hypospermatogenesis. An aberrant protamine substitution is also resultant effect of high histone methylation concentration that adversely affects quality of sperm. Protamines found in spermatozoa might undergo alkylation as a result of oxidative imbalance with results effect by impairment in spermatogenesis via chromatin condensation. Therefore, a lower concentration of nuclear protamine in the spermatozoa indicates that such is vulnerable to oxidative damage owing to easy accessibility DNA [97, 98]. Acetylation of histone is also a vital epigenetic regulation mechanism that could be impaired by oxidative stress in oocyte formation or spermatogenesis.

Several studies have reported the interconnection of altered chromatin remodeling and sperm DNA damage initiation [99].

5. Conclusion

Oxidative stress as a result of accumulated free radicals adversely affects the reproductive process in human by posing a negative impact on both the male and female gametes and rate of development of the embryo. Age also adversely effects oxidative stress and DNA methylation with a resultant effect in epigenetic disorders transmission by the elderly to their offspring. This chapter discussed the interplay among epigenetics, reproductive aging and oxidative stress and provided a broader overview of the molecular basis of reproductive aging associated with ROS production and the consequential effects on the epigenetic machinery in both male and female. This in essence give an overall understanding on the molecular processes associated with reproductive aging and diseases that may arise due to aging for a better approach in ameliorating or managing the physiological process.

From the foregoing, it is crystal clear that aging is attended by several physiological and cellular alterations. The role of oxidative stress in the furtherance of these alterations is profound. Oxidative stress increases with age due to a weakened antioxidant defense system and results in profound consequences. The epigenetic alterations that come with reproductive aging were discussed. However, a literature search shows that more work needs to be done on the influence of reproductive aging on histone modifications in sperm and oocytes.

Conflict of interest

The authors do not have any conflict of interest to declare.

Author details

Olalekan Bukunmi Ogunro[1*] and Oladimeji Taiwo Babatunde[2]

1 Department of Biological Sciences, Reproductive and Endocrinology, Toxicology, and Bioinformatics Research Laboratory, KolaDaisi University, Ibadan, Oyo State, Nigeria

2 Department of Chemical Sciences, Crown-Hill University, Eiyenkorin, Kwara State, Nigeria

*Address all correspondence to: olalekanbukunmi@gmail.com

References

[1] Pizzino G, Irrera N, Cucinotta M, Pallio G, Mannino F, Arcoraci V, et al. Oxidative stress: Harms and benefits for human Health. Oxidative Medicine and Cellular Longevity. 2017;**2017**:1-12

[2] Agarwal A, Aponte-Mellado A, Premkumar BJ, Shaman A, Gupta S. The effects of oxidative stress on female reproduction: A review. Reproductive Biology and Endocrinology. 2012;**10**(1):1

[3] Agarwal A, Virk G, Ong C, du Plessis SS. Effect of oxidative stress on male reproduction. The World Journal of Men's Health. 2014;**32**(1):1

[4] Huang M, Wu Q, Jiang ZH. Epigenetic Alterations under Oxidative Stress in Stem Cells. Oxidative Medicine and Cellular Longevity. 2022;**2022**:1-11

[5] Sharma P, Ghanghas P, Kaushal N, Kaur J, Kaur P. Epigenetics and oxidative stress: A twin-edged sword in spermatogenesis. Andrologia. 2019;**51**(11):1-10

[6] Chamani IJ, Keefe DL. Epigenetics and female reproductive aging. Frontiers in Endocrinology. 2019;**10**(7):1-7

[7] Ben Maamar M, Nilsson EE, Skinner MK. Epigenetic transgenerational inheritance, gametogenesis and germline development†. Biology of Reproduction. 2021;**105**(3):570-592

[8] Ostan R, Monti D, Gueresi P, Bussolotto M, Franceschi C, Baggio G. Gender, aging and longevity in humans: An update of an intriguing/neglected scenario paving the way to a gender-specific medicine. Clinical Science. 2016;**130**(19):1711-1725

[9] Hägg S, Jylhävä J. Sex differences in biological aging with a focus on human studies. eLife. 2021;**10**:1-27

[10] Richards G, Browne WV. Prenatal testosterone and sexually differentiated childhood play preferences: A meta-analysis of amniotic fluid studies. Current Psychology. 2022;(0123456789):1-14

[11] Gunes S, Hekim GNT, Arslan MA, Asci R. Effects of aging on the male reproductive system. Journal of Assisted Reproduction and Genetics. 2016;**33**(4):441-454

[12] Dutta S, Sengupta P, Slama P, Roychoudhury S. Oxidative stress, testicular inflammatory pathways, and male reproduction. International Journal of Molecular Sciences. 2021;**22**(18):1-20

[13] Tyagi V, Scordo M, Yoon RS, Liporace FA, Greene LW. Revisiting the role of testosterone: Are we missing something? Reviews in urology. 2017;**19**(1):16-24

[14] Jung JH, Kim HW, Narayan V, Reddy B, Walsh T, Yu HS, et al. Testosterone supplementation in men with sexual dysfunction. Cochrane Database of Systematic Reviews. 2018;**2018**(7):1-20

[15] Andersen AN, Wohlfahrt J, Christens P, Olsen J, Melbye M, Nybo A. General practice linkage study. BMJ. 2000;**320**(7):1708-1712

[16] Keegan J, Parva M, Finnegan M, Gerson A, Belden M. Addiction in pregnancy. Journal of Addictive Diseases. 2010;**29**(2):175-191

[17] Animasahun VJ, Chapman HJ. Psychosocial health challenges of the elderly in Nigeria: A narrative review. African Health Sciences. 2017;**17**(2):575-583

[18] Mikhael S, Punjala-Patel A, Gavrilova-Jordan L. Hypothalamic-pituitary-ovarian axis disorders impacting female fertility. Biomedicine. 2019;**7**(1):1-9

[19] Nordenström A, Ahmed SF, van den Akker E, Blair J, Bonomi M, Brachet C, et al. Pubertal induction and transition to adult sex hormone replacement in patients with congenital pituitary or gonadal reproductive hormone deficiency: An Endo-ERN clinical practice guideline. European Journal of Endocrinology. 2022;**186**(6):G9-G49

[20] Basaran M, Kosif R, Bayar U, Civelek B. Characteristics of external genitalia in pre- and postmenopausal women. Climacteric. 2008;**11**(5):416-421

[21] Thornton K, Chervenak J, Neal-Perry G. Menopause and sexuality. Endocrinology and Metabolism Clinics of North America. 2015;**44**(3):649-661

[22] Scavello I, Maseroli E, Di Stasi V, Vignozzi L. Sexual health in menopause. Medicina (Lithuania). 2019;**55**(9):1-18

[23] Almouzni G, Cedar H. Maintenance of epigenetic information. Cold Spring Harbor Perspectives in Biology. 2016;**8**(5):1-23

[24] Felsenfeld G. A brief history of epigenetics. Cold Spring Harbor Perspectives in Biology. 2014;**6**(1):1-10

[25] Moosavi A, Ardekani AM. Role of epigenetics in biology and human diseases. Iranian Biomedical Journal. 2016;**20**(5):246-258

[26] Heyn H, Li N, Ferreira HJ, Moran S, Pisano DG, Gomez A, et al. Distinct DNA methylomes of newborns and centenarians. Proceedings of the National Academy of Sciences of the United States of America. 2012;**109**(26):10522-10527

[27] McCartney DL, Stevenson AJ, Hillary RF, Walker RM, Bermingham ML, Morris SW, et al. Epigenetic signatures of starting and stopping smoking. eBioMedicine. 2018;**37**:214-220

[28] Pal S, Tyler J. Epigenetics and aging. Science Advances. 2016;**2**(7):1413-1433

[29] Soriano-Tárraga C, Jiménez-Conde J, Roquer J. Epigenetics and aging. Handbook of Nutrition, Diet, and Epigenetics. 2019;**2**(7):1413-1433

[30] Kumar S, Chinnusamy V, Mohapatra T. Epigenetics of modified DNA bases: 5-Methylcytosine and beyond. Frontiers in Genetics. 2018;**9**(12):1-14

[31] Miller J, PA G. The role of DNA methylation and histone modifications in transcriptional regulation in humans. Sub-Cellular Biochemistry. 2013;**61**:239-317

[32] Wotten D, Pemberton L, Meriiill-Schools J. SUMO and chromatin remodeling. In: Wilson V, editor. Wilson, V (eds). SUMO Regulation of Cellular Processes. Springer: Cham; 2017. pp. 171-184

[33] Tushir-Singh J, Bhatnagar S. In vitro assay to study histone ubiquitination during transcriptional regulation. Methods in Molecular Biology. 2017;**1507**(1):235-244

[34] Magaña-Acosta M, Valadez-Graham V. Chromatin remodelers in the 3D nuclear compartment. Frontiers in Genetics. 2020;**11**(11):1-22

[35] Carlson SM, Gozani O. Nonhistone lysine methylation in the regulation of cancer pathways. Cold Spring Harbor Perspectives in Medicine. 2016;**6**(11):1-16

[36] Moore AM, Frohwirth L, Miller E. Male reproductive control of women who have experienced intimate partner violence in the United States.

Social Science & Medicine. 2010;**70**(11):1737-1744

[37] Chen B, Chan W. The de novo DNA methyltransferase DNMT3A in development and cancer. Epigenetics. 2014;**9**(5):669-677

[38] Yue M, Fu X, Zhou G. Abnormal DNA methylation in oocytes could be associated with a decrease in reproductive potential in old mice. Journal of assisted reproduction and genetics. 2012;**29**(7):643-650

[39] Kawai K, Harada T, Ishikawa T, Sugiyama R, Kawamura T, Yoshida A, et al. Parental age and gene expression profiles in individual human blastocysts. Scientific Reports. 2018;**8**(1):1-10

[40] Zhang W, Xu J. DNA methyltransferases and their roles in tumorigenesis. Biomarker Research. 2017;**5**(1):1-8

[41] Sendžikaitė G, Kelsey G. The role and mechanisms of DNA methylation in the oocyte. Essays in Biochemistry. 2019;**63**(6):691-705

[42] Fedoriw A, Mugford J, Magnuson T. Genomic Imprinting and Epigenetic Control of Development. Cold Spring Harb Perspect Biology. 2012;**4**(7):1-16

[43] Ge ZJ, Schatten H, Zhang CL, Sun QY. Oocyte ageing and epigenetics. Reproduction. 2015;**149**(3):R103-R114

[44] Messerschmidt DM, Knowles BB, Solter D. DNA methylation dynamics during epigenetic reprogramming in the germline and preimplantation embryos. Genes and Development. 2014;**28**(8):812-828

[45] Rengaraj D. Female Germ Cell Development in Chickens and Humans: The Chicken Oocyte Enriched Genes Convergent and Divergent with the Human Oocyte. nternational journal of molecular sciences. 2022;**23**(19):1-17

[46] Yin L, Zhang Y, Lv P, He W, Wu Y, Liu A, et al. Insufficient maintenance DNA methylation is associated with abnormal embryonic development. BMC Medicine. 2012;**10**(1):26

[47] Breton-Larrivée M, Elder E, McGraw S. DNA methylation, environmental exposures and early embryo development. Animal Reproduction. 2019;**16**(3):465-474

[48] Kohli RM, Zhang Y. TET enzymes, TDG and the dynamics of DNA demethylation. Nature. 2013;**502**(7472):472-479

[49] Qian Y, Tu J, Tang NLS, Kong GWS, Chung JPW, Chan WY, et al. Dynamic changes of DNA epigenetic marks in mouse oocytes during natural and accelerated aging. International Journal of Biochemistry and Cell Biology. 2015;**67**:121-127

[50] Akiyama T, Nagata M, Aoki F. Inadequate histone deacetylation during oocyte meiosis causes aneuploidy and embryo death in mice. Proceedings of the National Academy of Sciences of the United States of America. 2006;**103**(19):7339-7344

[51] Manosalva I, González A. Aging alters histone H4 acetylation and CDC2A in mouse germinal vesicle stage oocytes. Biology of Reproduction. 2009;**81**(6):1164-1171

[52] Wang N, Satirapod C, Ohguchi Y, Park ES, Woods DC, Tilly JL. Genetic studies in mice directly link oocytes produced during adulthood to ovarian function and natural fertility. Scientific Reports. 2017;**7**(1):1-16

[53] Moghadam ARE, Moghadam MT, Hemadi M, Saki G. Oocyte quality and

aging. Jornal Brasileiro de Reproducao Assistida. 2022;**26**(1):105-122

[54] Demond H, Trapphoff T, Dankert D, Heiligentag M, Grümmer R, Horsthemke B, et al. Preovulatory aging in vivo and in vitro affects maturation rates, abundance of selected proteins, histone methylation pattern and spindle integrity in murine oocytes. PLoS One. 2016;**11**(9):1-18

[55] Grøndahl ML, Yding Andersen C, Bogstad J, Nielsen FC, Meinertz H, Borup R. Gene expression profiles of single human mature oocytes in relation to age. Human Reproduction. 2010;**25**(4):957-968

[56] Jenkins TG, Aston KI, Carrell DT. Sperm epigenetics and aging. Translational Andrology and Urology. 2018;**7**(Suppl 3):S328-S335

[57] Milekic MH, Xin Y, O'Donnell A, Kumar KK, Bradley-Moore M, Malaspina D, et al. Age-related sperm DNA methylation changes are transmitted to offspring and associated with abnormal behavior and dysregulated gene expression. Molecular Psychiatry. 2015;**20**(8):995-1001

[58] Kobayashi N, Okae H, Hiura H, Chiba H, Shirakata Y, Hara K, et al. Genome-scale assessment of age-related DNA methylation changes in mouse spermatozoa. PLoS One. 2016;**11**(11):1-13

[59] Oakes CC, Smiraglia DJ, Plass C, Trasler JM, Robaire B. Aging results in hypermethylation of ribosomal DNA in sperm and liver of male rats. Proceedings of the National Academy of Sciences of the United States of America. 2003;**100**(4):1775-1780

[60] Tatehana M, Kimura R, Mochizuki K, Inada H, Osumi N. Comprehensive histochemical profiles of histone modification in male germline cells during meiosis and spermiogenesis: Comparison of young and aged testes in mice. PLoS One. 2020;**15**(4):1-20

[61] Xie K, Ryan DP, Pearson BL, Henzel KS, Neff F, Vidal RO, et al. Epigenetic alterations in longevity regulators, reduced life span, and exacerbated aging-related pathology in old father offspring mice. Proceedings of the National Academy of Sciences of the United States of America. 2018;**115**(10):E2348-E2357

[62] Jan SZ, Hamer G, Repping S, de Rooij DG, Van Pelt AMM, Vormer TL. Molecular control of rodent spermatogenesis. Biochimica et Biophysica Acta - Molecular Basis of Disease. 2012;**1822**(12):1838-1850

[63] Smith LB, Walker WH. The regulation of spermatogenesis by androgens. Seminars in Cell and Developmental Biology. 2014;**30**:2-13

[64] Anifandis G, Messini C, Dafopoulos K, Sotiriou S, Messinis I. Molecular and cellular mechanisms of sperm-oocyte interactions opinions relative to in vitro fertilization (IVF). International Journal of Molecular Sciences. 2014;**15**(7):12972-12997

[65] Hirose M, Honda A, Fulka H, Tamura-Nakano M, Matoba S, Tomishima T, et al. Acrosin is essential for sperm penetration through the zona pellucida in hamsters. Proceedings of the National Academy of Sciences of the United States of America. 2020;**117**(5):2513-2518

[66] Maine EM. Gametogenesis. Brenner's Encyclopedia of Genetics: Second Edition. 2013;**2**(meiosis I):154-156

[67] Griswold MD. Spermatogenesis: The commitment to meiosis. Physiological Reviews. 2016;**96**(1):1-17

[68] Zidi-Jrah I, Hajlaoui A, Mougou-Zerelli S, Kammoun M, Meniaoui I, Sallem A, et al. Relationship between sperm aneuploidy, sperm DNA integrity, chromatin packaging, traditional semen parameters, and recurrent pregnancy loss presented at the 17th world congress on in vitro fertilization, Tunis, Tunisia, on September 4-7, 2013. Fertility and Sterility [Internet]. 2016;**105**(1):58-64

[69] Johnson GD, Lalancette C, Linnemann AK, Leduc F, Boissonneault G, Krawetz SA. The sperm nucleus: Chromatin, RNA, and the nuclear matrix. Reproduction. 2011;**141**(1):21-36

[70] Zou C, Mallampalli RK. Regulation of histone modifying enzymes by the ubiquitin-proteasome system. Biochimica et Biophysica Acta - Molecular Cell Research. 2014;**1843**(4):694-702

[71] Amor H, Zeyad A, Bakry MS, Bosilah AMH, Ben AH, Hammadeh ME. Protamine ratio as predictor of the fertility potential of sperm by couple undergoing ICSI. International Journal of Women's Health and Reproduction Sciences. 2018;**6**(4):400-409

[72] Elango K, Kumaresan A, Talluri TR, Raval K, Paul N, Samuel E, et al. Impact of sperm protamine on semen quality and fertility. Reproductive Healthcare and Medicine. 2022;**3**(5):1-9

[73] Oliva R. Protamines and male infertility. Human Reproduction Update. 2006;**12**(4):417-435

[74] Eun SH, Gan Q, Chen X. Epigenetic regulation of germ cell differentiation. Current Opinion in Cell Biology. 2010;**22**(6):737-743

[75] Kane AE, Sinclair DA. Epigenetic changes during aging and their reprogramming potential. Critical Reviews in Biochemistry and Molecular Biology. 2019;**54**(1):61-83

[76] Vatner SF, Zhang J, Oydanich M, Berkman T, Naftalovich R, Vatner DE. Healthful aging mediated by inhibition of oxidative stress. Ageing Research Reviews [Internet]. 2020;**64**:101194

[77] Sasaki H, Hamatani T, Kamijo S, Iwai M, Kobanawa M, Ogawa S, et al. Impact of oxidative stress on age-associated decline in oocyte developmental competence. Frontiers in Endocrinology. 2019;**10**(N):1-7

[78] Wang L, Tang J, Wang L, Tan F, Song H, Zhou J, et al. Oxidative stress in oocyte aging and female reproduction. Journal of Cellular Physiology. 2021;**236**(12):7966-7983

[79] Ben-Meir A, Burstein E, Borrego-Alvarez A, Chong J, Wong E, Yavorska T, et al. Coenzyme Q10 restores oocyte mitochondrial function and fertility during reproductive aging. Aging Cell. 2015;**14**(5):887-895

[80] Mihalas BP, Redgrove KA, McLaughlin EA, Nixon B. Molecular mechanisms responsible for increased vulnerability of the ageing oocyte to oxidative damage. Oxidative Medicine and Cellular Longevity. 2017;**2017**:1-22

[81] Perkins AT, Das TM, Panzera LC, Bickel SE. Oxidative stress in oocytes during midprophase induces premature loss of cohesion and chromosome segregation errors. Proceedings of the National Academy of Sciences of the United States of America. 2016;**113**(44):E6823-E6830

[82] Yusuf I, Emokpae MA. Association between a marker of sperm dna damage and sperm indices in infertile

males in Benin City, Nigeria: A cross-sectional study. International Journal of Reproductive BioMedicine. 2021;**19**(2):137-146

[83] Nguyen-Powanda P, Robaire B. Oxidative stress and reproductive function in the aging male. Biology. 2020;**9**(9):1-15

[84] Diamanti-Kandarakis E, Dattilo M, Macut D, Duntas L, Gonos ES, Goulis DG, et al. Mechanisms in endocrinology: Aging and anti-aging: A combo- endocrinology overview. European Journal of Endocrinology. 2017;**176**(6):R283-R308

[85] Giorgi C, Marchi S, Simoes ICM, Ren Z, Morciano G, Perrone M, et al. Mitochondria and reactive oxygen species in aging and age-related diseases. International Review of Cell and Molecular Biology. 2018;**340**:209-344

[86] Nilsson MI, Tarnopolsky MA. Mitochondria and aging—The role of exercise as a countermeasure. Biology. 2019;**8**(2):1-18

[87] Sharifi-Rad M, Anil Kumar NV, Zucca P, Varoni EM, Dini L, Panzarini E, et al. Lifestyle, oxidative stress, and antioxidants: Back and forth in the pathophysiology of chronic diseases. Frontiers in Physiology. 2020;**11**:1-21

[88] Vargas-Mendoza N, Angeles-Valencia M, Morales-González Á, Madrigal-SantillánEO,Morales-MartínezM, Madrigal-Bujaidar E, et al. Oxidative stress, mitochondrial function and adaptation to exercise: New perspectives in nutrition. Life. 2021;**11**(11):1-24

[89] Liguori I, Russo G, Curcio F, Bulli G, Aran L, Della-Morte D, et al. Oxidative stress, aging, and diseases. Clinical Interventions in Aging. 2018;**13**:757-772

[90] Guillaumet-Adkins A, Yañez Y, Peris-Diaz MD, Calabria I, Palanca-Ballester C, Sandoval J. Epigenetics and oxidative stress in aging. Oxidative Medicine and Cellular Longevity. 2017;**2017**:1-14

[91] Kandlur A, Satyamoorthy K, Gangadharan G. Oxidative stress in cognitive and epigenetic aging: A retrospective glance. Frontiers in Molecular Neuroscience. 2020;**13**(41):1-8

[92] Etchegaray JP, Mostoslavsky R. Interplay between metabolism and epigenetics: A nuclear adaptation to environmental changes. Molecular Cell. 2016;**62**(5):695-711

[93] Kreuz S, Fischle W. Oxidative stress and oxygen radicals/oxidative stress in health and disease. Epigenomics. 2016;**8**(6):843-862

[94] Tahiliani M, Koh KP, Shen Y, Pastor WA, Bandukwala H, Brudno Y, et al. Conversion of 5-methylcytosine to 5-hydroxymethylcytosine in mammalian DNA by MLL partner TET1. Science. 2009;**324**(5929):930-935

[95] Kafer GR, Li X, Horii T, Suetake I, Tajima S, Hatada I, et al. 5-Hydroxymethylcytosine Marks sites of DNA damage and promotes genome stability. Cell Reports. 2016;**14**(6):1283-1292

[96] Lv S, Wen H, Shan X, Li J, Wu Y, Yu X, et al. Loss of KMT2D induces prostate cancer ROS-mediated DNA damage by suppressing the enhancer activity and DNA binding of antioxidant transcription factor FOXO3. Epigenetics. 2019;**14**(12):1194-1208

[97] Cescon M, Chianese R, Tavares RS. Environmental impact on male (in) fertility via epigenetic route. Journal of Clinical Medicine. 2020;**9**(8):1-33

[98] Wang T, Gao H, Li W, Liu C. Essential role of histone replacement and modifications in male fertility. Frontiers in Genetics. 2019;**10**:1-15

[99] Bunkar N, Pathak N, Lohiya NK, Mishra PK. Epigenetics: A key paradigm in reproductive health. Clinical and Experimental Reproductive Medicine. 2016;**43**(2):59-81